D1330513

Divining a Digital Future

Divining a Digital Future

Mess and Mythology in Ubiquitous Computing

Paul Dourish and Genevieve Bell

The MIT Press
Cambridge, Massachusetts
London, England

For information about special quantity discounts, please email special_sales@mitpress.mit.edu

This book was set in Stone Sans and Stone Serif by Toppan Best-set Premedia Limited. Printed and bound in the United States of America.

Library of Congress Cataloging-in-Publication Data
Dourish, Paul.
Divining a digital future : mess and mythology in ubiquitous computing / Paul Dourish and Genevieve Bell.
p. cm.
Includes bibliographical references and index.
ISBN 978-0-262-01555-4 (hardcover : alk. paper)
1. Ubiquitous computing. 2. Computer networks—Social aspects—Forecasting.
I. Bell, Genevieve. II. Title.
QA76.5915.D68 2011
303.48′33—dc22
2010038387

10 9 8 7 6 5 4 3 2

Contents

Preface vii
Acknowledgments ix

1 Introduction: The Myth and Mess of Ubiquitous Computing 1

I 7

2 Contextualizing Ubiquitous Computing 9

3 Making Room for the Social and Cultural 45

4 A Role for Ethnography: Methodology and Theory 61

II 91

5 What Lies Beneath 95

6 Mobility and Urbanism 117

7 Rethinking Privacy 137

8 Domesticity and Its Discontents 161

III 185

9 Reimagining Ubiquitous Computing: A Conclusion 187

References 211
Index 245

Preface

We wrote the proposal for this book in a stark modernist hotel room in Geneva, while attending a new media conference there. In between conference sessions, extravagant French-influenced desserts, and a memorable trip to the International Museum of the Reformation, we sketched out a plan for how it might take shape. We knew from the outset that we wanted to critically interrogate the idea of "ubiquitous computing" (or "ubicomp"). It was a project that we had been circling around in various publications and talks for several years, and the time seemed ripe to take a more comprehensive look.

Much of this project's distinctiveness and whatever success it achieves relies on the interdisciplinary nature of our collaboration. Dourish is a computer scientist whose work lies at the intersection of computer science and social science; Bell is a cultural anthropologist with a primary concern in information technology as a site of cultural production along with the consequences for technology innovation and diffusion. Our intellectual and personal trajectories are complicated. We have each spent time, in various guises—as a child, worker, student, professor, and researcher—in a range of significant hubs—such as the Australian National University, Stanford University, Bryn Mawr College, Cambridge, University College London, Silicon Valley, Rank Xerox EuroPARC, Xerox PARC, Apple, and Intel's Architecture Lab and Corporate Technology Group. Raised in Australia and educated in the United States, Bell is currently directing a new Interaction and Experience Research laboratory at Intel Corporation. Raised in Scotland, and educated in Scotland and England, Dourish now runs an interdisciplinary research program at the University of California, Irvine.

Our collaboration reflects our belief that any satisfactory account of contemporary computational practice must be deeply grounded in cultural and specific settings yet must also take the nature of computational devices seriously, not black boxing them, but instead engaging with both the technology and practice of information technology design. The material in this book, then, draws on the body of empirical and conceptual work that we have undertaken, individually and together, over the last six years, across a wide range of related topics, domains of inquiry, and regions of the world. Our work has generated a series of articles in journals and other publications over the last three or four years; these writings form the basis of the book and are elaborated, developed, connected, and reframed in terms of a broader set of conceptual and methodological themes.

The book itself is divided into three sections. The first section spells out the terrain of ubicomp with a particular focus on culture and ethnography as theoretical as well as methodological stances. The second one provides a sequence of thematic explorations of dominant narratives within ubicomp. The third section provides a conclusion, proposing directions for future exploration and study.

We wrote this book over a long period, and it has gone through multiple instantiations and a number of working titles. But the one thing that has been a constant is the cover image. It first appeared in Australia in 1985. Produced by Redback Graphix, an activist design house run by Michael Callaghan from Wollongong, "Bush Radio" sought to promote a joint venture—bringing radio in local languages to the aboriginal communities of central Australia—between the Australian Broadcasting Corporation and the Central Australian Aboriginal Media Association. The image blends contemporary representations of radio waves and broadcast ranges with a map of the Northern Territory, while simultaneously evoking the more traditional images of Aboriginal art and country. Here, the "future" is understood by reference to the present and the ever present. It is just such forms of hybridity, the reappropriations of technical realities, around which this book turns.

Acknowledgments

Although we both attended a conference organized by Lucy Suchman and her colleagues in 1999, we managed not to meet until an Intel-sponsored conference in Oregon in September 2003. Other conferences followed, and over time an intellectual connection bloomed into an ongoing collaboration that has seen us writing papers together for nearly seven years. This book is a natural outgrowth and result of that process, but it would never have come to print without a great deal of assistance, input, and support.

Some specific material for this book has been drawn from written pieces and projects that we have undertaken in collaboration with others, including Ken Anderson, Arianna Bassoli, Kirsten Boehner, Johanna Brewer, Simon Cole, Rogerio de Paula, Eric Kabisch, Scott Mainwaring, Karen Martin, Phoebe Sengers, Jennifer Terry, and Amanda Williams.

We have benefited greatly from a sustaining network of colleagues who, through conversation, collaboration, and critique, have helped shape the ideas we present here.

At Intel, this includes Ken Anderson, Ashwini Asokan, Maria Bezaitis, Sue Faulkner, Anthony LaMarca, Scott Mainwaring, Jay Melican, Dawn Nafus, Mike Payne, Tony Salvador, John Sherry, and Alex Zafiroglu.

At the University of California, Irvine, this includes Tom Boellstorff, Simon Cole, Beatriz da Costa, Martha Feldman, David Theo Goldberg, Gillian Hayes, Garnet Hertz, Mimi Ito, Crista Lopes, Liz Losh, George Marcus, Gloria Mark, Bill Maurer, Melissa Mazmanian, Bonnie Nardi, Robert Nideffer, Don Patterson, Simon Penny, Kavita Philip, Mark Poster, David Redmiles, and Jennifer Terry.

Beyond the immediate boundaries of our own institutions, this includes Mark Ackerman, Julian Bleecker, Geoffrey Bowker, John Seely Brown, Matthew Chalmers, Kate Crawford, Michael Curry, Keith Edwards, Bill Gaver, Gerard Goggin, Melissa Gregg, Beki Grinter, Steve Harrison, Larrisa Hjorth, Kia Höök, Ann Light, Peter Lunenfeld, John McCarthy, Malcolm

McCullough, Christena Nippert-Eng, Eric Paulos, Yvonne Rogers, Christine Satchell, Phoebe Sengers, Brian Cantwell Smith, Lucy Suchman, Nina Wakeford, and Peter Wright.

At the outset of our collaboration, we found ourselves writing about infrastructure, and so we sent a draft of our first paper on the topic to Leigh Star, whose thinking and writing on infrastructure and messiness were as inspiring to us then as they are now. Leigh's generous and thoughtful engagements with us, around that paper and in the years that followed, were invaluable. Leigh passed away a few weeks before this manuscript was completed. Her presence is hugely missed both personally and intellectually.

Dourish's students and postdocs, past and present, have made—and continue to make—key contributions through their research, questions, and refreshing refusal to take anything at face value: Johanna Brewer, Judy Chen, Marisa Cohn, Rogerio de Paula, Danyel Fisher, Lilly Irani, Charlotte Lee, Silvia Lindtner, Madhu Reddy, Jennifer Rode, Irina Shklovski, Janet Vertesi, and Amanda Williams.

We thank our current institutions for their forbearance, support, and engagement: UC Irvine's Department of Informatics as well as the Laboratory for Ubiquitous Computing and Interaction, and Intel Corporation (in particular, the Corporate Technology Group's Peoples and Practice Research along with the Digital Home Group's User Experience Group). Aspects of this work were also supported in part by the National Science Foundation under awards 0133749, 0205724, 0326105, 0524033, 0527729, 0712890, 0838499, 0838601, and 0917401, and by grants from the Intel Research Council and Intel's Digital Home Group. Early material was developed while Paul was on sabbatical at Stanford University, where Terry Winograd, Scott Klemmer, and their students were his welcoming and engaging hosts.

We should also thank the MIT Press, and in particular Doug Sery for his commitment to this project and to us, willingness to take this book on when it was just a title and short abstract, and remarkable patience as it developed.

On a personal note, we also acknowledge the support of Brian David Johnson, Katrina Jungnickel, Heather Masterton, Josh Rohrbach, and Melinda Stelzer.

1 Introduction: The Myth and Mess of Ubiquitous Computing

Ubiquitous computing names the third wave in computing, just now beginning. First were mainframes, each shared by lots of people. Now we are in the personal computing era, person and machine staring uneasily at each other across the desktop. Next comes ubiquitous computing, or the age of calm technology, when technology recedes into the background of our lives.

—Mark Weiser, "The Computer for the 21st Century"

In Palo Alto, California, on Coyote Hill Road, in sight of the foothills of the coastal range, the Xerox Corporation runs a research and development center. Xerox founded its Palo Alto Research Center, or PARC as it is known, in 1970, and it has gone on to be a significant node in the cultural geography of Silicon Valley.[1] PARC is a leading research center and the site where, famously, a small group of researchers in the 1970s invented many of the elements of the contemporary personal computing environment—personal workstations with graphical user interfaces with overlapping windows, mice, local area networking, digital typography and document production, and more. PARC also helped create new stories about how technology would fit into the world; the personal computer, the graphic user interface, the paperless office, and ubiquitous computing are arguably the most enduring ones. The stories, or organizing visions, told in the pages of publications like the *New York Times* and *Scientific American*, were aimed at both technical and nontechnical audiences. For insiders, these visions created the opportunities for new research projects and publications; for the general public, they were something more. They prefaced new realities and new promises, and in so doing they echoed previous technology visions—the electrical age, the radio age, the television age, and even the atomic age.

1. After operating since 1970 as Xerox PARC, PARC was spun off as an independent but wholly owned subsidiary, PARC, Inc., in 2002.

Like those earlier technovisions, PARC's technotales would also become myths: they would create a way to make sense of the future that appeared simultaneously magically but also manageably. That these myths emanated from the center of Silicon Valley gave them a sense of inevitability as well. After all, if smart engineers and computer scientists say this is our future, then surely it will be true. This kind of rhetorical positioning also meant that to be skeptical of such visions was to be seen as against progress, a Luddite or worse. And like all good myths, there would be heroes, seemingly impossible tasks, perils, pitfalls, and dangers, and of course, in the end, glory.

In the late 1980s and early 1990s, a team of researchers at PARC, led by computer scientist Mark Weiser, found itself in a world shaped by two different yet increasingly convergent mythical stories. The team operated in a research culture framed by the "personal computer" story about the transformation of massive mainframe industrial computation machinery into something smaller, more intimate, and with the power to change human relations with technology and each other. It was also the first days of a new era, the "information age," also arguably mythical, where binary code would replace physical labor and information would trump mechanization as a driving economic force. Inhabiting a world very much bookended by these two stories—the personal computer and the information age—Weiser and his team, following early PARC researcher Alan Kay's injunction to predict the future by inventing it, staked their own claim in the technomythscape.

In talks, publications, and hallway conversations, a story about the next future of computation and also the next stage of the future of humanity emerged. This tale coalesced, in 1991, around the notion of ubiquitous computing (ubicomp)—a vision, as articulated by Weiser, that made sense of the information age while suggesting that personal computing had not gone far enough.

Weiser argued that the first era of computing had been that of mainframes—large, centralized computers used by hundreds or thousands of people. The second era, personal computing, was characterized by "a computer on everyday desktop," a world in which computational resources were deployed on a personal level. In the third era, ubiquitous computing, he contended that computational devices would be small and powerful enough to be worn, carried, or embedded in the world around us—in doors and tables, the fabric of clothes and buildings, and the objects of everyday life.

Computing technology, in this ubicomp vision, would be everywhere, anticipatory, and far more practical—it would be useful as well as extraordinary. In this way, it was a familiar formulation—a technological breakthrough that would, when realized properly, change social relations, social order, and daily life, creating new possibilities, both commercial and cultural. In the meantime, ubicomp was a useful organizing principle for industrial and academic research, conferences, journal articles, and papers along with prototypes, test beds, and experimentation. It has come to be broadly recognized in academic, commercial, and government settings worldwide as one of the key agendas for information technology research. And it has held sway, in a range of sites and guises, for more than twenty years. Influencing more than two generations of scholars, it has become a foundational story, a technomyth, in computer science and allied fields and as a result has shaped the kinds of technologies that have been made and also made possible.

Writing toward the end of the twentieth century, the pioneers in ubicomp research tried to anticipate the impacts and applications of their technologies decades into the future. That time, of course, is now, and many aspects of their vision have been realized, at least from a technological perspective. Weiser anticipated a world suffused with information technology, in which daily life might bring some people into contact with many, interconnected digital devices, large and small. For many people, in many parts of the world, this is indeed a fair characterization, but it only goes so far. Important considerations were unexamined or unexpected by the early researchers, from the widespread use of mobile communications technology in the developing world to the impact of location-based services on how Japanese teens interact, the emergence of new forms of political engagement online, or the need for legislation to curb our use of distracting devices while driving.

In this book, we examine the process of "divining a digital future." "Divining" has multiple meanings here. Most immediately, we consciously evoke the notion of divination—the complex and somewhat mystical process of inquiring into future events. We are struck, relatedly, by the link to the kinds of things that people do with divining rods—looking to uncover what lies hidden from immediate sight. At the same time, the notion of the divine—a search for transcendental phenomena, and a process by which some truths are found to lie beyond the realm of the mundane—is also implicated in the contemporary practice of conjuring technological futures. This is the broad landscape, but our particular attention is more locally to the domain of ubicomp in which we are both ourselves situated. Taking ubicomp to be at once a technological and an imaginative effort, we explore the vision that has driven the ubicomp research agenda and the contemporary practices that have emerged. Drawing on cross-cultural investigations of technology adoption, we argue for developing a "ubiquitous computing of the present" that takes the messiness of everyday life as a central theme.

Our goal is to understand the mythology of ubicomp. When we talk in terms of myths, we do not mean to suggest that ubicomp is somehow false or mistaken. We instead want to direct attention toward the ideas that animate and drive ubicomp forward, in much the same way that myths provide human cultures with ways of understanding the world and celebrating their values. As Vincent Mosco (2004, 3) notes:

> Useful as it is to recognize the lie in the myth, it is important to state at the outset that myths mean more than falsehoods or cons; indeed, they matter greatly. Myths are stories that animate individuals and societies by providing paths to transcendence that lift people out of the banality of everyday life. They offer an entrance to another reality; a reality once characterized by the promise of the sublime.

The myths we want to examine, then, are the stories that motivate and celebrate the development of the ubicomp agenda. They are the ideas that give it shape and meaning. They are ideas about what technology can do for people, the places it will go, and the needs it will address. While we might not often see technology in mythical terms, it is a useful strategy to uncover the ideas that shape our technological world—the ideas about human action that spurred early researchers in cybernetics and artificial intelligence (Hayles 1999; Pickering 2010), the cold war rhetoric that drove the development of digital computing (Edwards 1996), the notions of politics and community that inflected the discourse of contemporary web technologies (Coyne 1999; Mosco 2004), or the visions of life and death at work in the artificial life community (Helmreich 1998).

Alongside the myth, there is the mess—the practical reality of ubicomp day to day. We do not use the term "mess" pejoratively; we rather like the mess (as anyone would be able to see who glanced at the space where we sit writing these words). When we talk of the mess, we want to suggest that the practice of any technology in the world is never quite as simple, straightforward, or idealized as it is imagined to be. For any of the infrastructures of daily life—the electricity system, the water system, telephony, digital networking, or the rest—the mess is never far away. Lift the cover, peer behind the panels, or look underneath the floor, and you will find a maze of cables, connectors, and infrastructural components, clips, clamps, and duct tape. Push further, and you will also encounter the regulatory authorities who authorize interventions and certify qualified individuals, committees that resolve conflicting demands in the process of setting standards, governments that set policy, bureaucrats who implement it, marketers who shape our views of the role of the infrastructure in our lives, and more. Mess is always nearby.

"Mess" refers, too, to the way that technological realities are always contested. No single idea holds about what technologies are and what they do. Though many have tried, attempts to reduce this complexity to a single reading are at best unsatisfactory; as Andrew Pickering (2010, 33) observes, "Ontological monotheism is not turning out to be a pretty sight." So partly our concerns with mess highlight not just an interest in "how things could have been different" but rather how they already are different among the different groups, places, contexts, and circuits that characterize contemporary ubicomp.

This book, then, is about ubicomp. It is about the stories that have been told, and all the stories that haven't been. It is about the research that has been done, and the research that should be done. It is about what computer science has been, at the intersection of daily life and computational technology, and what it could be. It is then a book about the myth of ubicomp and its messy reality and, by necessity, about the tensions between those two very different vantage points. As such, there are many things that this book is not. It is not an ethnographic account of the ubicomp community, though surely such an account is necessary. It is not a recitation of current ubicomp experiments and a reporting out of results; we leave that to other forums. It is also not an easy read or a quick fix for ubicomp. We are concerned instead with offering a thorough and rigorous critique. In so doing, we hope to open up ubicomp to a larger audience and to make room for a far more diverse set of practitioners, collaborators, and engagements.

As a project, a "ubiquitous computing of the present" would necessarily reach beyond computer science as a disciplinary foundation. Information technology is certainly a major component here, and indeed we find those projects that ignore the materiality and practical consequences of information technology as unsatisfying as those unable to see beyond it. In attempting to understand what ubicomp is today, however, we need to understand it not just technically but also culturally, socially, politically, and economically. Often, this means starting off by understanding it historically—understanding where it came from and what kinds of ideas and hopes contributed to its development. This will be our starting point here.

So at the same time, this project is something of an interdisciplinary experiment, and one fraught with not a little danger. As a socioculturally inclined computer scientist and a technologically inclined anthropologist, we have each always been oriented toward unconventional modes of analysis within our own disciplines. Working and writing together over the last few years, we have been able to join our voices with those of many others who have been working over many years, from different places and in different ways, to fashion a new disciplinary perspective on information technology and its workings in the world. Some amount of this project is thus a tentative exploration of alternate configurations of disciplinary and scholarly practice. A ubicomp of the present is both our topic and an exemplar of disciplinary hybridity that we find intriguing.

However, first things first: in this case it means ubicomp and its first stirrings in Palo Alto.

I

2 Contextualizing Ubiquitous Computing

"The most profound technologies," wrote Weiser (1991, 78) in the opening of his classic *Scientific American* article that laid the foundations for a research program in ubicomp, "are those that disappear. They weave themselves into the fabric of everyday life until they are indistinguishable from it." Weiser was a computer scientist with wide-ranging interests. He had conducted significant research in programming systems, including advanced techniques such as program slicing and automated storage management ("garbage collection"), and was at the time director of the Computer Science Laboratory (CSL) at Xerox PARC. Weiser's influential article set out to provide the basis for a new paradigm and to document PARC's recent work in developing it. When Weiser explained that "my colleagues and I at the Xerox Palo Alto Research Center think that the idea of a 'personal' computer itself is misplaced," his words had especial salience, since these were the very people who had developed the idea of the personal computer in the first place.

The article is partly a manifesto and partly a progress report. As a manifesto, it follows in a long technological tradition (Brate 2002); it paints a picture of an alternative to the personal computing paradigm and extrapolates technological trends in order to present this alternative as both necessary and inevitable. As a progress report, it documents a range of contemporary research developments—primarily although not solely at PARC—that constitute steps along the path.

It is structured in several sections. Beginning with its resonant and widely quoted argument about technologies that "disappear," it suggests that a significant problem has arisen with the personal computing paradigm, which forces people to spend too much time "interacting with" their computers rather than using them to get things done. "The state of the art," Weiser remarks, "is perhaps analogous to the period when scribes had to know as much about making ink or baking clay as they did about writing."

The assertion that poorly designed computer systems have a tendency to get in the way and interfere with the tasks that people are trying to carry out was far from a new one; this, after all, was the central argument motivating research on human-computer interaction (HCI) and its efforts to focus attention on the user interface and experience. For social scientists Geoff Cooper and John Bowers (1995), in their essay on the disciplinary rhetorics of HCI, different formulations of this problem created the opportunity to wrest control of technological developments from computer scientists and create spaces where cognitive scientists and social scientists could bring their own expertise to bear. Weiser, however, takes a different tack—one that is fascinating with respect to interdisciplinary engagement. He makes two arguments simultaneously. First, he contends that the underlying problem is not simply one of the user interface but rather about the very conception of computational devices. This is an issue that will require significant engineering efforts and pose tremendous computer science challenges. Yet, at the same time, it is a problem that draws even more widely on other disciplines than HCI has traditionally done. Weiser invokes ecological psychologists and phenomenological philosophers as well as cognitive scientists and computer scientists in his attempts to reframe the problem. This is an engineering challenge, to be sure, he maintains, but it will not be business or research as usual.

In point of fact, Weiser's ubicomp was informed by a number of nontechnical impulses and sources. Ubicomp was in no small way a polemic about the perceived failure of the personal computer to deliver meaningful value to human beings. In a retrospective published eight years after the *Scientific American* article and just weeks after Weiser's death, John Seely Brown (then the chief technology officer of PARC) wrote: "We wanted to put computing back in its place, to reposition it into the environmental background, to concentrate on human-to-human interfaces and less on human-to-computer ones" (Weiser, Gold, and Brown 1999, 694). In the telling and retelling of the ubicomp story, discussions with colleagues at Xerox and beyond who were critical of the traditional conceptions of computation, interaction, and practice embedded in computer system design play an important part. Weiser and his colleagues shared the PARC facility with the Work Practice and Technology Group—a team led by anthropologist Lucy Suchman (1987, 1999, 2007; Blomberg, Suchman, and Trigg 1997) that took solid aim at issues of HCI, privileging the human side of the interaction while bringing a different set of methodological and theoretical impulses to PARC.

Writing of Suchman and her team, Weiser notes,

> To some of the technologists at PARC, myself included, their observations led toward thinking less about particular features of a computer—such as random access memory and number of pixels or megahertz—and much more about the detailed situational use of the technology. In particular, how were computers embedded within the complex social framework of daily activity, and how did they interplay with the rest of our densely woven physical environment (also known as "the real world")? (Weiser, Gold, and Brown 1999, 693)

In other words, sociological and ethnographic accounts of work practice and interaction had begun to suggest alternatives to traditional "cognitivist" accounts of interaction with computer systems and had emphasized the importance of looking at the systems of practice within which HCI was embedded (e.g., Orr 1996; Suchman 1999, 2007; Weiser, Gold, and Brown 1999). If the technical trends that Weiser examined looked to a day when computation could move "off the desktop," these alternative models of HCI suggested that this was also where the orderliness and "calm" of interaction was to be located as well as achieved (Weiser and Brown 1997). While our contemporary computational lives might scarcely be described as calm or even orderly (Rogers 2006), we might nonetheless conclude that in a world of cell phones, wireless computer networks, MP3 players, digital cameras, social networking websites, and virtual worlds, Weiser shrewdly anticipated technological trends.

The vision that Weiser laid out in 1991 replaced the traditional personal computer, desktop, or laptop with a range of small computational devices distributed through the everyday world and embedded into it. These devices operate at different scales. Weiser analogized them to different elements of the traditional office environment—devices on the scale of sticky notes, pads of paper, and whiteboards—connected via wireless networking technologies so that they can communicate and interoperate. In his approach, the computational experience moved away from the desktop and into the everyday world and was distributed across a range of devices, each specialized to particular sorts of tasks.

After outlining the overall vision, Weiser's article then moves on to report progress to date. As Weiser sets out the components of his ubicomp environment, he also presents the prototype systems developed by his colleagues at Xerox PARC, describing both their design and early experiences of their use. The effect of this portrayal is twofold. First, it shows the reality of the vision; no scientific pipe dream, Weiser's account of an alternative to the desktop computing paradigm is already becoming reality, as the lavishly illustrated article demonstrates. Second and relatedly, it begins to shape an argument for the future of ubicomp as inevitable; it is already on its way, delivered by the people who brought you the personal computer.

There is one further crucial resonance in the way that the progress report is provided. When the personal computing vision was emerging at PARC in the early 1970s, one of the laboratory's strategies was to develop and widely deploy its technologies within its own environment, and to live with and use them daily. PARC researcher Kay is famous for his observation that the best way to predict the future is to invent it; a corollary is that the best way to understand the future is to do your best to create a local approximation and try to use it day to day. Much of Xerox's success in developing its vision of personal computing had come from the researchers' ability to engage in this form of time-machine research. Weiser's article detailed the way that researchers were doing this once more. So while ubicomp constituted a break from the past, the research tools and strategies that would realize it would be those that had succeeded the last time around.

The inevitability of the new vision is further bolstered in the next section of his article, in which Weiser lays out trends in the development of key technologies. Here again, a familiar technological trope is deployed. In 1965, semiconductor engineer and Intel cofounder Gordon Moore first observed that the number of transistors that could be placed on a wafer of silicon had for several years been doubling every eighteen months. Moore did not formulate this as a prediction, but as it continued to hold over the next few years, others began to refer to it as "Moore's law," which came to be seen as a predictive model of technological development. Arguably, as it became widely accepted, it became somewhat self-fulfilling, as engineers addressed themselves to the problem of just how the next doubling in feature density might be achieved on schedule. In his article, Weiser invokes Moore's law to anticipate changes in computational power density and similarly extrapolates trends in storage and networking capacity. Weiser's predictions are reasonably accurate and even somewhat conservative. More interestingly, though, they further suggest the inevitability of his account; one has the sense that he is telling the reader, "This is coming, so you had better be ready." Computers are coming to the people.

This is perhaps clearest in the next section of his article, where with some trepidation, Weiser sketches a fictional scenario of what it would be like to live in a world of ubicomp. His protagonist here is Sal, a mother and Silicon Valley executive whose work and home lives are suffused with the sorts of technologies as well as interaction motifs that Weiser has been outlining. When introducing this segment, Weiser notes the difficulties of posing these sorts of fictional predictions. Like most futurists, too, he finds that he needs to present his image of an alternate form of life within a familiar frame, so as to highlight the specific transformations being posited. Our attention is thus inevitably drawn to the assumptions about the world in which the new technological arrangements will be situated. It is notable, for instance, that Sal works in a place quite like Xerox's own research facility—a place where she interacts informally with colleagues, manages her own professional obligations, works regularly from home, and spends her working hours engaged in a combination of design and writing. More broadly, Sal finds herself in the sort of workplace that, as in today's world, is already suffused with information technology—the kind of office environment into which the desktop personal computer was introduced and that it has now thoroughly colonized. Ubicomp does not therefore create new user constituencies, nor does it challenge the assumptions around which the workplace is organized; it instead augments these environments, replacing desktop computers with new devices (and creating a whole new demand for upgrading the technological environment). Sal, Weiser seems to say, is "us." Of course, the particularities of Sal's life providing such fertile ground for Weiser's imagined computational world are the same ones that locate ubicomp in a U.S. middle-class frame.

Nonetheless, Weiser's article struck a chord. It presented a new model for human-computer interaction—one that offered new challenges and motivated both design and engineering activities. For people working in hardware design, it opened up new opportunities by showing how small, specialized devices could play a role in larger systems. For researchers working in distributed systems, it created new challenges of scale as well as turning the spotlight on the interactional consequences of design decisions. Ubicomp, then, became an animating vision for a wide range of computer science research areas; rather than a new topic in itself, it supplied a new approach to a wide range of research topics, some old and some new.

While the pursuit of personal computing had been a unifying force for the laboratory in the 1970s, the 1980s had been a decade of consolidation that had also been marked by the fragmentation of collective attention into the traditional subdisciplines of computer science—mathematical theory, networks, distributed systems, programming languages, hardware design, graphics and imaging, and so forth. In reinvigorating CSL's research agenda, Weiser deliberately sought an approach in which each of these areas would have important contributions to make and that might in turn become a labwide program. Consequently, when the ubicomp research agenda spread beyond the walls of CSL and PARC, it did so in a way that created a new nexus for interdisciplinary engagement. Rooted in traditional areas of computer science research, especially the design of digital networks, mobile and embedded hardware design, distributed systems (research into software systems whose operation requires the coordination of several different network-linked components), and software architecture (the study of approaches to the decomposition and structuring of large-scale software systems), ubicomp emerged as a research area that drew on but extended each of these existing domains. Further, as ubicomp research moved beyond CSL, it came into more sustained connection with other ongoing areas of research, including HCI and computer-supported cooperative work (CSCW).

Weiser's articulation of a novel research program around ubicomp was in these respects remarkably successful. In other ways, though, it was less so. As a research domain, ubicomp has remained largely rooted in computer science, where it cuts across traditional affiliations and focus areas. Yet such technological developments clearly have had massive ramifications not simply in their economic import and associated transformations of the workplace but culturally too, in the ways in which the ideas of information, flows, and networks have become animating narratives for contemporary accounts of life and society (Woolgar 2002). So to the extent that ubicomp provides both a platform for encounters between people and technology, on the one hand, and an animating vision for technologically mediated interaction, on the other, it places itself (not always consciously) also into a broader disciplinary conversation with science and technology studies, sociocultural anthropology, and media and cultural studies. Weiser himself understood this. A slide from his keynote talk at the Association for Computing Machinery (ACM) Symposium on User Interface, Systems, and Technologies in 1994 is especially telling. Titled "Building Invisible Interfaces" it reads, "Start from arts and humanities: philosophy, phenomenology, anthropology, psychology, postmodernism, sociology of science, feminist criticism, your own experience." Under this, written in boldface type, is the following: "This is the most important part of the talk. You may not get it on first hearing. Patience" (Weiser 1994, slide 10).

Ubicomp after Weiser

Weiser died unexpectedly in 1999 after a brief battle with cancer. He was just forty-six years old and still a highly productive researcher and research manager at the time. In the years between the publication of the *Scientific*

American article and his death, ubicomp continued to gain traction at PARC and found willing participants elsewhere. In Weiser's lab, ubicomp deployments still build on the work that he articulated in 1991 (Want et al. 1995). Elsewhere, context-aware computing research at EuroPARC and Georgia Tech involved both small-scale experiments (e.g., Lamming and Flynn 1994; Long et al. 1996) and medium-scale deployments (e.g., Want et al. 1992). There was also emerging research at IBM's Tokyo lab around cell phone applications and mobile banking (Scheter 2000). For the most part, ubicomp, as it was becoming known as, was centered on working prototypes, test beds, and experimentation.

By the time of Weiser's death, however, competing narratives were taking shape at other technology companies: ambient intelligence at Philips (Aarts and Marzano 2002; Aarts, Harwig, and Schuurmans 2001; Zelkha and Epstein 1998) and later adopted by the European Commission's Information Society and Technology Advisory Group (IST Advisory Group 2001; Riva et al. 2005), pervasive computing at IBM (Ark and Selker 1999), and proactive computing at Intel (Tennenhouse 2000). All these technology-future stories shared an orientation to the rapid growth in technology deployments, the growing importance of the Internet, and the challenges that consumers, government, and industry faced in negotiating a world rich in digital devices. They differed in their research agendas, however. For instance, pervasive computing at IBM had an explicit focus on mobile devices and phones, in addition to cars and homes, while proactive computing at Intel articulated an interest in machine learning, data processing, and algorithms as well as personal experiences with digital technologies.

During this same period of the late 1990s and early 2000s, at various U.S. and British universities, research that centered on ubicomp agendas was also commencing: the Aware Home Research Initiative at Georgia Tech (Abowd et al. 2000), Project Oxygen (Rudolph 2001) and later the Internet of Things (Gershenfeld, Krikorian, and Cohen 2004) at MIT, and Project Aura at Carnegie Mellon (Garlan et al. 2002) all bore a strong connection to Weiser's earlier articulations of ubicomp. They were centered for the most part in computer science departments. In the United Kingdom, by contrast, Equator, a multisite project across eight British universities, was formulated around the integration of physical and digital interaction and had a strong interdisciplinary component drawing on sociologists, ethnomethodologists, and social psychologists as well as computer scientists.

This latter project is a significant one because, perhaps more than many others, it starts to explicitly marry technological experimentation with social and critical theory. After all, in addition to his charge to "start from . . . your own experience," Weiser had also suggested that ubicomp would find much inspiration and potentially useful grounding in a range of academic disciplines and theoretical approaches. Prior to Weiser's death, this had happened rarely. While Weiser might have hoped that future research in ubicomp would be thoroughly grounded in postmodern analysis and feminist critical theory, among others, these have been minority contributions at best to the ubicomp research literature, as measured by publications in the leading conferences such as Ubicomp and Pervasive Computing, both international research meetings sponsored by the major professional societies in the area.

In many ways, Weiser's first *Scientific American* article was all about establishing a sort of authority, fixing a technical narrative, and perhaps even making the myth that would be ubicomp. Over the years, there have been a number of articles tracing the development of ubicomp, some with a more critical eye than others (Weiser, Gold, and Brown 1999; Abowd and Mynatt 2000; Suchman 2002; Rogers 2006; Bell and Dourish 2007; Greenfield 2006; McCullough 2004). Establishing the history of the making of ubicomp, then, is an important discursive practice, and one worth examining further.

One exemplar is an influential paper published in the ACM's *Transactions on Computer-Human Interaction,* in a special issue assessing the state and prospects of research in various areas of HCI research (Abowd and Mynatt 2000). Titled "Charting Past, Present, and Future Research in Ubiquitous Computing," it explicitly addresses the progress in the years since Weiser's article. The authors, Gregory Abowd and Elizabeth Mynatt, are both members of the faculty in the College of Computing at the Georgia Institute of Technology, a leading research institution in the area of ubicomp. Each has a long track record of influential research in the area. Mynatt worked with Weiser at PARC for several years in the mid-1990s, having been hired as one of the first HCI researchers at CSL.

Given the scope of research on ubicomp, Abowd and Mynatt (2000, 30) limit themselves to a particular set of areas—"*natural interfaces* that facilitate a richer variety of communication capabilities between humans and computation," "*context-aware* [applications], adapting their behavior based on information sensed from the physical and computational environment," and "applications [that] automate the *capture* of live experiences and provide flexible and universal *access* to those experiences later on." In each of these areas, their paper catalogs research to date and outlines key challenges. Finally, they introduce a new thematic area of concern for researchers of ubicomp, or what they label "everyday computing," which in their view "promotes informal and unstructured activities typical of much of our everyday lives. . . . Familiar examples are orchestrating tasks, communicating with family and friends, and managing information" (ibid.).

What we see at work in Abowd and Mynatt's paper is a complicated tension between two pressures, which in Kuhnian terms we might call the pressures toward a "paradigm shift" and "normal science." On the side of the paradigm shift, we see their attempts to retain some of the visionary elements of Weiser's proposal. They hold on to the sense that (although now some ten years since the earliest explorations) what is on offer in ubicomp is a radical departure from the computational status quo. Indeed, the very figuring of the everyday as a new domain for research attention, and a site that presents new and underexamined challenges, reflects Weiser's own efforts to look beyond the boundaries of traditional computer science. On the other hand, the pressure toward normal science is seen in two ways. It is conveyed in the way that the visionary expressions of ubicomp are here broken down into a series of engineering challenges, amenable to the traditional and incremental approaches of computer science design. Yet at the same time, it is manifest in the list of paradigmatic examples of the new domain, which are remarkably pedestrian. Orchestrating tasks and managing information are problems to which computer scientists had been applying themselves since the dawn of the field. They are, for instance, topics toward which Vannevar Bush's attention was clearly directed in his classic, widely read, hugely influential article "As We May Think" (1945). The tension between the impulse to radically reconfigure the discipline and the need to systematically organize one's work within that discipline's frame is clearly on display.

One reason for this, of course, is that ubicomp is no longer up for grabs in quite the sense that it was for Weiser. Weiser had a blank canvas to fill; Abowd and Mynatt wrote in a world already hemmed in by conferences, journals, courses, publications, thesis requirements, professional practices, working groups, corporations, and products for which the term "ubicomp" was already invested with meaning. Visions of technology reflect the times and places where they originate.

In her paper "Located Accountabilities in Technology Production," Suchman (2002) builds on work in feminist epistemology and connects it to the practices of technological design. She outlines three positions that frame the working relations of design: the view from nowhere, detached intimacy, and located accountabilities. Suchman directs us toward the importance of the contexts within which design engagements take place, and the power relations at work in the encounters between those framed as technology providers and those framed as potential consumers, recognizing that designers operate within a range of relations—not only with their "users" but also with their corporate colleagues, professional communities, and more. With that in mind, it is perhaps not surprising that PARC's implementations of Weiser's vision look the way they do; when ubicomp research is sponsored by Xerox, the technologies and systems that emerge are likely those that operate within the sorts of environments into which Xerox traditionally delivers solutions (office environments within a dominantly Western commercial context). It is perhaps less clear why the visions pursued by academic researchers should similarly focus on these sorts of environments, although the regular traffic back and forth between commercial and academic research institutions, the shared contexts of professional research communities, and the agenda-setting influence of industry within the academy certainly all play significant roles (Barley, Meyer, and Cash 1988; Slaughter et al. 2002).

Certainly, it should be noted that Abowd and Mynatt, separately and together, have been among the foremost proponents of a shift in ubicomp away from workplace applications and toward those in domestic spaces (e.g., Kidd et al. 1999; Rodden and Benford 2003). One of the critiques of this approach, however, has been that although the research setting has changed, much of the context remains the same—the context of technology and media consumption, the instrumental accounts of information technology use, and the problems of infrastructure and maintenance (for a further discussion of this, see chapter 8).

As we noted, Abowd and Mynatt's focus is of necessity selective and, naturally enough, focused on areas in which they are themselves experts. What is particularly intriguing in relation to the discussion above is that connections to the social sciences and humanities are almost entirely absent, although one might imagine that those domains would have some light to cast on "natural" interfaces, context, and the capture, representation, and recall of human experience—the three topics they selected.

By contrast, the social sciences and their evolving position within computer science research is much more central in an article by Yvonne Rogers (2006) that assesses the state and prospects of ubicomp research. Rogers writes of "moving on from Weiser's vision," but the vision she has in mind is not simply that expressed in his original article; it is also one articulated in a related, later piece (cowritten with PARC director Brown) in which, using the term "calm computing," Weiser somewhat reframed the ubicomp vision in terms of computational experiences that could live in or move to the periphery of human attention, rather than those that obstinately and obstreperously occupied the center. Rogers writes too in the context of more recent trends in HCI that have seen an increasing interest in art and design, entertainment, games, and arenas of human experience with computation and computational devices beyond the office and "workaday world" (Moran and Anderson 1990).

Rogers's argument centers on the issue of engagement. Essentially, she maintains that a notion of ubiquitous or calm computing that withdraws into the periphery fails to create the kinds of expansive, playful, and engaging experiences that promote human participation in new domains. In this she envisions an emancipatory and democratic information technology, such as technology that involves people in public debates around issues of science and governance concerning climate change, environmental pollution, and health care, while simultaneously drawing on research in the social sciences about learning, participation, motivation, and behavior change.

Rogers frames her contention as "moving on" from Weiser's original outline of the scope and opportunities of ubicomp, and indeed the examples she draws on and domains she addresses herself to certainly go well beyond the working settings with which Weiser was concerned (even when Weiser addressed himself to domestic life, in the scenario of Sal, he told a story about working from home and the expansion of the working sphere into domestic space—a topic about which we have more to say in chapter 8). Still, there are ways in which Rogers's call signals a return to programmatic elements of Weiser's work that have become less prominent in subsequent years. In her account of alternative domains for applications of ubicomp and assessments of its effectiveness drawn from other disciplines, Rogers echoes Weiser's central assertion that in "getting computing out of the way," Weiser's goal is not to make it physically invisible but rather to have it play a role in agendas that originate elsewhere. Recall, for example, his appraisal that the "most important" thing to know about how to go about designing ubicomp experiences was to start from arts and humanities.

So while Weiser's domains and examples—reflecting, again, the positions and sites from which he was writing—are ones focused on the world of work, his strategy is one quite in line with Rogers's argument. Rogers's call is an important one, given the way that some of these concerns have faded in the intervening years; but while she might advocate going beyond some of Weiser's considerations, in other instances she is pushing forward aspects of his own agenda.

Envisioning the Future

Today, ubicomp research is characterized primarily by a concern with potential future computational worlds—what we call the "proximate future" (Bell and Dourish 2007).[1] This notion of research by future envisionment has been a feature of ubicomp discourse and reasoning since its earliest days; Weiser's foundational article (1991) is even titled "The Computer for the 21st Century"—an explicit look toward a possible future. Rhetorically, Weiser situates the research activities that he describes there as initial steps on a path of technological development inspired by an explicit vision of possible future relationships between people, practice, and technology. Although much of his article depicts a research program already under way along with some of the early results that it produced, the dominant theme is the twin challenges of anticipating future trends and meeting future needs.

Weiser's article was doubly influential. Not only did it articulate a research agenda that many have embraced, it also set a rhetorical tone that many have adopted. The same concern with technological futures thus continues to feature in the ways in which ubicomp research agendas are framed, and in which technological advances are motivated and measured. Ubicomp is essentially defined by its visions of a technological future. Often, this is taken directly from Weiser's own work; Weiser's foundational articles are cited at every annual Ubiquitous Computing conference (which began in 2001, a decade after the original publication), sometimes by as many as 40 percent of the papers.[2] Even in cases in which Weiser's own vision is not a driving factor, the idea that ubicomp research is exploring prototypes of tomorrow's everyday technology and experience is a pervasive one.

Whether we look at this from the perspective of the late 1980s, when the dominant vision of ubicomp was being formulated, or in 2010, as we write this, what we want to keep sharply in focus the relationship between ubicomp research and technological practice. Clearly, to the extent that ubicomp was a visionary proposal when first articulated, it told a story of

1. We are inspired by *Memory Practices in the Sciences* (Bowker 2006, 33), which contrasts the proximate future (social time) with deep reality (natural time).
2. The rate for 2001 was 40 percent. Between 2002 and 2006, the rate varied between 17 and 27 percent. In 2007 and 2008, the rates were 7 and 10 percent, respectively—a lower but still remarkable number of papers to ground themselves in a single vision more than fifteen years later.

an as-yet-unattainable technological future, but that story is one firmly rooted in its own times. Such visions, after all, are interesting not just for what they say about the future but also for what they say about the present. This seems to be particularly the case when it comes to normative social relationships. Visions of the future, such as those of the World's Fair (Rydell 1993), Disney's Tomorrowland (Bukatman 1991), or most popular science fiction (Dourish and Bell forthcoming; Penley 1997) have provided a useful analytic lens for considering how the problems of today are perceived, framed, and understood. Yet when we adopt a contemporary perspective, we see a different relationship between research and practice at work. Many aspects of the original ubicomp research vision are mundane realities for people throughout the world; what was once research imaginary is now commonplace and unremarkable. The interplay between research and practice is a troublesome one, since a field that is successful runs the risk of becoming irrelevant. What this opens up as a topic of inquiry, then, is how new futures get to be imagined and incorporated into a research agenda such as ubicomp, and what kind of work has to be done to mark out past triumphs, current problems, and future opportunities.

In this chapter, we are concerned with the balance between past, present, and future embedded in conventional discourses about ubicomp. We are particularly interested in the central conundrum posed by the fact that Weiser's vision of the future is not only by this point an old one but also an extremely North American one. The role of technology in everyday life is, in the early twenty-first century, already quite different than it was when Weiser wrote in the late 1980s; among other things, it is now explicitly acknowledged to have remarkable cultural variation. As we noted, PARC famously pursued a policy of time-machine research, devoting its considerable financial and intellectual resources to creating simultaneous prototypes and simulations of future computing environments; at this stage in the game, though, conventional computing platforms have vastly outstripped even the "futuristic" environment that Weiser's laboratory was capable of building. Yet his original framing of the ubicomp vision still occupies an important place in much of the discourse of ubicomp research. And while it is tempting to speculate on the role that his early death has played in the durability of the ubicomp vision, the questions we want to ask instead are as follows: First, how should we understand the relationship between ubicomp's envisioned future and our everyday present? Second, what influence does this have on contemporary ubicomp research? Third, what motivates and explains the remarkable persistence as well as centrality of Weiser's vision? In seeking to answer these questions, we have arrived at three framing points that run through both this chapter and the book.

First, the centrality of ubicomp's proximate future continually places its achievements out of reach, while at the same time blinding us to current practice. By concentrating on the future just around the corner, ubicomp renders contemporary practice (outside research sites and "living labs") by definition irrelevant or at the least already outmoded. Arguably, however, ubicomp is already here; it simply has not taken the form that we originally envisaged and continue to conjure in our visions of tomorrow. We draw on two holistic case studies in this chapter to substantiate this point. In sharp contrast to Weiser's Sal scenario, these short illustrations, focusing on Singapore and South Korea, are grounded in current lived realities, accounting for government regulations, economic, political, and historical forces, and cultural practices. Singapore and South Korea offer examples of strongly technologically oriented nation-states, geographically and culturally removed from Weiser's ubicomp mythology.

Second, the framing of ubicomp as something yet to be achieved allows researchers and technologists to absolve themselves of responsibilities for the present; the problems of ubicomp are framed as implementation issues that are essentially someone else's problem, to be cleaned up afterward as part of the broad march of technology or to be solved by savant children (Facer and Furlong 2001; Facer et al. 2003). The future framing allows us to assume that certain problems will simply disappear of their own accord; questions of usability, regulation, resistance, adoption barriers, sociotechnical backlashes, and other concerns are erased. By looking at case studies from countries in which the vision of ubicomp has played out differently than Weiser foresaw, we draw attention to the complex settings within which ubicomp is always already embedded.

Third, the seamlessly interconnected world of future scenarios is at best a misleading vision and at worst a downright dangerous one. In many parts of the world, cultural and social practices privilege disconnection, seams, and discrete distinct realms of activity and action (Bell 2006a; Jungnickel and Bell 2008). Industry and government too benefit from noninteroperability around telecommunications and broadcast standards, not to mention content regulation and protection. While homogeneity and an erasure of differentiation is a common feature of future environments, the practice is inevitably considerably messier, and perhaps dealing with the messiness of everyday life should be a central element of ubicomp's research agenda. We illustrate the work involved in continually producing alignments between technological opportunities and social realities.

The Problem of the Proximate Future

The dominant tense of ubicomp writing is that of the proximate future. That is, motivations and frames are often written not merely in the future tense, describing events and settings to come, but portray a proximate future, one just around the corner. The proximate future is invoked in observations that "Internet adoption levels will shortly reach . . . ," "We are entering a period when . . . ," "New technological opportunities are emerging that . . . ," or "Mobile phones are becoming the dominant form of. . . ." A brief perusal of the proceedings of recent relevant conferences affirms the pervasive sense of the proximate future; of the 108 papers comprising the annual International Conference on Ubiquitous Computing proceedings between 2001 and 2008, almost 40 percent are oriented toward a proximate (and inevitable) technological future (e.g., from only the 2004 proceedings, Beckman, Consolvo, and LaMarca 2004; Borriello et al. 2004; Hull, Clayton, and Melamed 2004; Kindberg, Sellen, and Geelhoed 2004; Masui, Tsukada, and Siio 2004; Patel and Abowd 2004; Truong, Huang, and Abowd 2004; Vermuri et al. 2004).

It may be that subsequent ubicomp writing has adopted not only Weiser's technological vision but aspects of his formulation of how this vision will come to pass too. Certainly, this collective envisioning of a future saturated with technology has been a defining characteristic of ubicomp research. What is perhaps most interesting is that ubicomp research has generally shared not only in the notion of a technology-saturated future but also in the designations of what sorts of technologies in particular will saturate our future. In other words, these collective avowals of future potentialities are ways in which current activities are tied to the selfsame agenda that Weiser set out in the late 1980s. Citations to Weiser's article are in fact frequently phrased not so much as a "look backward" but rather as a collective "look forward"; that is, instead of saying "back in 1991, we thought that . . . ," they say "Just as Weiser suggested in 1991, we are soon to enter a world where. . . ."

Bruno Latour (1987) and others in the sociology of science have talked about the pragmatics of citation practice, noting that citing the ways in which one's own work depends on prior results and writings not only acknowledges intellectual debts but also builds defensible positions by aligning research with existing paradigms and traditions, and by enrolling others as tacit supporters of one's approach. Beyond the ways in which citation builds intellectual networks, however, it is worth stopping for a moment to think about the time dilations involved here. Weiser formulated his vision of ubicomp in the late 1980s (Weiser, Gold, and Brown 1999); although the *Scientific American* article was published in 1991, the ubicomp project was already well under way by that point, and indeed the 1991 article is not only a compelling vision of the future but also a progress report on the project of realizing it (complete with photographs of devices that had already been designed and built and reports on their use). Yet the invocation of Weiser's vision as one that we share and continue to prosecute neglects the significant difference between then and now along with the changing technosocial contexts.

So, for example, in 1989, when the ubicomp research agenda was being formulated at Xerox PARC, Intel introduced the 486DX processor, running at 25 megahertz (MHz). Apple introduced the Mac IIci, based on a 25-MHz MC68030 processor. It had yet to release a PowerBook laptop, but Apple also debuted its 16-pound, 16-MHz Macintosh Portable in 1989. In that same year, Sun introduced the Sparcstation-1, running at 20 MHz and with a maximum memory capacity of 64 MB. From a contemporary perspective, these would be poor "specs" for a portable MP3 player, never mind a scientific workstation. The figures for the telecommunications market are equally salutary. The first U.S. cellular telephony service had begun in 1983; by 1988, there were approximately 1.6 million U.S. subscribers. In Europe, the (now-dominant) global system for mobile (GSM) communication standard for mobile telephony was not demonstrated until 1991 (with the first network operator becoming active in 1992). In Japan, prior to government deregulation of mobile telephony, an average consumer could expect to pay $700 USD to sign up for service and another $300 per month in basic fees (Weinberg 1996).

In other words, today's technological landscape is radically different than that of the late 1980s when Weiser was outlining the ubicomp vision. It is a surprising idea that in the early twenty-first century, we are postulating much the same proximate future vision of ubicomp that motivated Weiser. Given that the last twenty years have seen such dramatic transformations of technological infrastructure worldwide, and that Moore's law suggests an eight-thousand-fold increase in computational performance, two questions immediately present themselves. First, why is our vision of the future still the same as Weiser's, and second, why has it not yet come to pass? Two possibilities present themselves.

The first possibility is that the ubicomp vision can never come to pass. The proximate future is a future infinitely postponed; when we are continually about to enter a new age, when we are continually anticipating what happens next, and when our attention is continually directed over the horizon, then by definition ubicomp is never about the here and now. Indeed, within this particular model of a technological future, it is hard to imagine how we could ever as a community say, "There. It is done."

The second possibility is that ubicomp already has come to pass. Clearly, few of us live in anything resembling Sal's world, and certainly not as described in the scenario outlined in Weiser's paper. But perhaps ubicomp is already here, somewhere in the world, and it has simply taken a different form than had been envisioned. Arguably, and as we will explore at more length below, our contemporary world—in which mobile computation and telephony are not just central aspects of Western commercial endeavors but also facets of everyday life in a range of different countries and cultures—is already one of ubicomp, albeit in unexpected forms.

These two possibilities are not in fact distinct options but rather two aspects of the same observation: that the vision of the future originally proposed by Weiser, and still motivating much research in ubicomp, is firmly entrenched in its own particular moments, locations, and cultural contexts—a vision as much of the past as of the future. From this perspective, the future that ubicomp has been attempting to build is not our own future; it is the 1989's future, or yesterday's tomorrows (Bell and Dourish 2007). Weiser envisioned the computer for the twenty-first century. Having entered that time period, what we should perhaps attend to is the computer of now.

Ubicomp Is Really about Messiness

In 1991, and in the intervening years until his death in 1999, Weiser and his colleague Brown made much of the notions of calm computing and "calm technology" (Weiser 1991; Weiser and Brown 1997). Although these concepts accompanied the vision of ubicomp and were frequently paired with it in publication, they have not had the same enduring legacy. It is nonetheless an interesting thread at which to pull. For in our collective vision of ubicomp's proximate future, the messiness of our local laboratory infrastructures (the nests of cables hidden in the dropped ceiling or behind the closet door; the jumble of Perl, Java, and Python code that precariously conspire to produce results in demos) is replaced by a clean, gleaming infrastructure seamlessly providing well-understood services—the epitome of calm computing. In practice, though, we see that infrastructures are continually visible and must be consciously attended to in the course of routine encounters with ubicomp, from the vagaries of network access to the structure of service billing. The critical property of this messy infrastructural regime in the everyday world is that it is most emphatically not a problem of living on the "bleeding edge," as it often is for research labs. Infrastructures remain messy after decades or centuries, as the user of any transit system from urban subways to international airlines can attest. The lesson of the real world of ubicomp is that we will always be assembling heterogeneous technologies to achieve individual and collective effects, and they will almost always be messy.

As we explore in more detail in chapter 5, Susan Leigh Star (1999, 2002) was an especially prominent advocate of the use of infrastructure as an analytic lens through which to consider the relationship between human action and technology. The crux of her approach is to look at infrastructure as a relational concept; an infrastructure is an infrastructure only from the perspective of specific peoples and technologies. To us as casual users, the sewer system is an urban infrastructure, ubiquitously available and uniform in its operation; to a city engineer, however, exposed to the daily practicalities and pragmatics of wastewater management and treatment, the sewer is not an infrastructure but rather a site of work, and not uniform but instead highly localized and variable.

Star's perspective is particularly instructive here, since in adopting infrastructure as a site of ethnographic inquiry, she also dispels the myth of infrastructure as quiescent and stable. Infrastructures must be actively maintained, and relationships to them must be continually negotiated. Scott Mainwaring, Michele Chang, and Ken Anderson (2004) provide a valuable examination of these issues for ubicomp by looking at different attitudes toward infrastructure and the relationship with infrastructure as itself a cultural production. So, for example, among their subjects, the rejection of infrastructure (be it commercial or technological) is itself a marker of certain forms of social life and even a way in which interpersonal relationships are managed (e.g., through a concern with "authenticity" in everyday life and interaction). Malcolm McCullough (2004) similarly explores the structure of everyday space as a confluence of infrastructural arrangements that overlap to produce effects that reach beyond each.

To put it another way, infrastructures are always already messy. The messiness that we experience in laboratory ubicomp infrastructures is not a property of prototype technologies, the bleeding edge, or pragmatic compromise; messiness is a property of infrastructure itself. Infrastructures are inherently uneven in their operation and availability. The notion of a seamless and uniform infrastructure is at best a chimera and at worst, to draw on an Aboriginal Australian cultural formulation, a *moolyewonk*—

a fearsome creature that might be invoked to steer people away from certain paths, places, or actions.

This messiness is important. Our suggestion that ubiquitous computing is already here, in the form of densely available computational and communication resources, is sometimes met with an objection that these technologies remain less than ubiquitous in the sense that Weiser suggested. Mobile telephony, after all, offers widespread coverage, but is neither truly ubiquitous nor truly seamless; incompatible standards, spotty regional coverage, and so forth, seem like obstacles that we must still overcome before the ubicomp vision can be realized. But postulating a seamless infrastructure is a strategy whereby the messy present can be ignored, although infrastructure is always unevenly distributed, always messy. An indefinitely postponed ubicomp future is one that need never take account of this complexity. Consider some examples.

Jojada Verrips and Birgit Meyer (2001) describe the complex networks of support and practice necessary to maintain cars in Ghana. Most vehicles on the road there were not designed for sub-Saharan Africa but rather are second- or thirdhand automobiles imported from Europe and pressed into new life. Verrips and Meyer's discussions of the travails of keeping a car on the road are framed by the radical departure from the European models of cars as commodity items and elements in a network of technological standardization. Initially surprised (indeed appalled) by the practices of customization, jury-rigging, and makeshift maintenance that are pervasive in Ghana, Verrips and Meyer gradually uncover an alternative infrastructure uniquely adapted to local needs and the problems of maintaining an engineering artifact outside its natural environment. The same vehicle, moved from Europe to Africa, is embedded in a dramatically different infrastructure as well as web of social values associated with mobility, reuse, exchange, commodification, craft, and so on.

In their work on information infrastructures, Geoffrey Bowker and Star (1999) discuss the International Classification of Diseases (ICD), a common infrastructure for the collection and comparison of mortality statistics worldwide. Like other boundary objects (Star and Griesemer 1989), though, the ICD is less a stable platform on which everyone can stand and more a means by which different interests, groups, concerns, and activities can be brought into temporary alignment. What seems like a straightforward process—the categorization of causes of death—is rife with complications, especially because of the uses to which the information will be put later. For example, consider the difficulties faced by AIDS researchers. AIDS deaths are typically the direct result of other infections that a patient suffers because of immune deficiency. Yet before a category allowed physicians to record death by AIDS, the condition was essentially "invisible" in the medical statistics, making it extremely difficult to mobilize support for research funding. Similarly, the different purposes to which the statistics, once gathered, might be put—such as public health actions, regulation of industries, allocation of resources, and more—result in a host of different pressures that shape the information infrastructure. Here the ICD's uniform structure hides the messiness of the practice beneath. For instance, the ICD coordinates multiple different interests including the needs of medical practitioners, legislators and regulatory agencies, researchers and funding agencies, pharmaceutical companies, and so forth. Furthermore, it is put to work in different ways in different parts of the world, reflecting not only different power relationships but also different cultural interpretations of disease.

A significant infrastructure issue for ubicomp endeavors is the problem of power—not MIPS per watt or even Foucauldian curbs on agency, but the electric supply that keeps our electronic world (quite literally) humming. World travelers are already familiar with the problems of varying voltage, frequency, and socket shapes, yet these forms of variability mask a more complex reality where, for example, in many parts of the developing world, the dominant infrastructure for power is based on car and truck batteries. It is not simply that these are improvised replacements for the "natural" arrangement of power distribution available in the West but rather as an alternative, indigenously appropriate, and thoroughly invisible infrastructure for power generation, albeit one quite radically different from our own. Ironically, current developments in what is known as "distributed generation" in the developed world (incorporating solar cells, fuel cells, and microturbine generators into a complex infrastructure in which production and consumption are more evenly distributed through the grid) are moving Western power infrastructures in directions closer, in some ways, to those we might see in the developing world.

Infrastructures, be they networks of car mechanics, medical categories, or power sources, are therefore never seamless in the ways in which they are put to work. They are sites of negotiation and contest, compromise and coordination, approximation and partial agreement. They are unevenly distributed and unevenly available. They are continually in flux and brought into local stability only through active engagement and coordination. Infrastructure itself is a relational property; it describes a relationship between technology, people, and practice. In this environment, then, thinking of infrastructure as stable, uniform, seamless, and universally available is clearly problematic. It is not merely a dream of a world not yet realized; it is a dream of a world that could never be realized.

Alternate Visions of Ubicomp

What might ubicomp already look like? With a few notable exceptions (e.g., Griswold et al. 2004; Bardram et al. 2006; Hayes et al. 2008), the available accounts of ubiquitous computing tend to be bounded, temporary deployments of hacked devices or nonscalable applications with target populations, and even studies of computing in daily life beyond the deceptively familiar test environments of the United States and Europe are remarkably rare in ubicomp research publications. Given the increasingly wide range of spaces and places richly populated with wireless and embedded technologies, there would seem to be a growing range of examples of ubicomp. Perhaps the challenge is that we do not recognize ubicomp when we see it, as mentioned earlier; perhaps it is because all the cultural, social, and political contexts within which technologies have transpired, and the contexts in which they been imagined, deployed, consumed, and resisted, are unfamiliar and thus unreadable.

In 2005, the South Korean Home Networking Industries Association launched a project called the Ubiquitous Dream Hall, in concert with South Korean governmental agendas around the role of new technologies in shaping Korean life, scientific discoveries, and economic engines. The hall itself fits within a long trajectory of smart home/smart office installations and has much that is familiar. Similarly, in 2008 the website that accompanies the project cites Weiser and offers glimpses into many facets of a recognizable ubicomp agenda, in which even the tropes are surprisingly familiar.[3] Its "digital life" scenarios acknowledge the complications of the social/work interfaces as well as what we really do in the bathroom (read newspapers), and there is a dizzying array of smart beds, sensing doorknobs, intelligent toilet doors, networked gym equipment, and anthropomorphic cleaning equipment along with the ever-present smart refrigerator ordering additional cartons of milk. That said, the website also departs from Weiser's vision and many of its contemporary instantiations: it lays claim to "ubiquitous" as verb, noun, object, and desired end-state. It promises a "vivid and emotionally satisfying future"—one

3. The South Korean Home Networking Industries Association's Ubiquitous Dream Hall Web site is available at <http://www.ubiquitousdream.or.kr> (accessed April 2008).

replete with images not often associated with the future, technology, or ubicomp. On a page called "U-Topia"—the "u" is for ubiquitous—the Korean Home Networking Industries Association links ubicomp to sulfur-crested cockatoos, parrot fish, curly-headed users sitting in noded apartment complexes that spring treelike from an undulating earth/ocean, satellite stars, an accordion-playing robot, hummingbird cellular phones, and battery-powered wishing wands (see figure 2.1).

It is probably safe to say that this is not ubicomp as Weiser envisioned or described it, but if we allow that this is indeed a vision of ubicomp, replete with embedded computation devices, smart and proactive networks, sensors and real-time feedback, what else might come into focus too? Looking beyond Sal's scenario and other standard visions of digital futures, we might find monitoring and restricting car traffic, centralized governmental control and regulation of infrastructure or content, dense urban environments, the extended collective public living, sensing urination in elevators, electronic consultations about feng shui, or ambient

Figure 2.1
An example of wallpaper for the Korean Home Network Industries Association Ubiquitous Dream Hall website.

displays of prayer times on a mosque wall. Where else might we see ubicomp already happening? Cairo, with its freshly deployed Wi-Fi network set to connect all the local mosques and create a single citywide call to prayer? Indonesia's e-mosque project? Singapore's real-time traffic control system?

A focus on current practice and the diversity of settings in which ubicomp is presently being put to work perhaps can help us avoid visions of ubicomp applications that are, in their own way, as misguided as 1950s' science fiction's speculations about twenty-first-century clothing and gender relations. In other words, by looking outside the research laboratory, we are looking at ubicomp as it is currently developing rather than as it is imagined. In these settings, we certainly see concerns with mobile devices, infrastructures for partial connectivity, and optimizing applications for devices with restricted input and restricted power. But we also see that the ubicomp agenda is one that is fundamentally tied to other important, though neglected, issues such as multigenerational living, high-density housing, public transit, religious observance, the practicalities of calling a cab, the politics of domesticity, and the spatialities of information access—the messiness of everyday practice.

Everyday Ubicomp: Singapore

Located at the tip of the Malay Peninsula, Singapore is a small, prosperous former British colony with a robust economy, technologically literate population, and reputation for strong government regulation of daily life (Huff 1994; Hui, Kiong, and Ser 1997; Koh and Ooi 2000; Murray and Perera 1996; Rengier 1987). Singapore is a diverse ethnic and cultural mix: more than 76 percent of Singapore's 4.5 million residents identify as Chinese, another 14 percent identify as Malay, slightly over 8 percent as Indian, and the rest are a mix of expatriates. Most Singaporeans are comfortably middle class, and the nation enjoys higher than average per capita household incomes (around US$48,900 per year) along with remarkable rates of home ownership (94 percent). Singapore is also one of the most wired countries in the world (Yew 2000), with a vast majority of homes connected to high-speed Internet. It is an example of ubicomp out of the lab, working. Yet Singaporeans do not appear to think of themselves as living on a cutting technology edge—cell phones, hot spots, and multimedia mobile messages are at this point naturalized parts of the local cultural landscape.

In 1992, six months after Weiser's article appeared in *Scientific American*, the Singaporean National Computer Board (1992) launched IT2000

Masterplan, the third in a series of ambitious technology road maps for its island nation-state. To achieve the plan's goals, the board deemed it necessary to develop a nationwide information infrastructure. This proposed infrastructure would connect up nearly every home, school, and workplace in Singapore, creating an "intelligent island" (ibid.). The island nation-state of Singapore is small—slightly less than 434-square miles, a landmass only three and half times that of Washington, DC—so making such a vision real seemed reasonable. As part of the broader IT2000 Masterplan, Singapore One Network for Everyone was launched in 1996, bringing together a range of public sector agencies to roll out a high-speed data network test bed across the island, which would in turn support a host of applications and services designed to enhance not only Singapore's economic position globally but also "improve the quality of life of its citizens" (Choo 1997). This plan was not without its technical challenges and delays (Ang, Zhou, and Jiang 2003), but it is safe to say now, a decade later, that the intelligent island vision has been realized. In 2006, the government launched its next ten-year plan, remaking the intelligent island as the Intelligent Nation 2015 (iN2015). The website of the Infocomm Development Authority of Singapore (IDA 2006) offers this introductory explanation of iN2015 (helpfully pronounced "in 2015"):

> In less than ten years, every single person and business in Singapore will find the world—and everyday life—transformed by technology. iN2015 is the blueprint to navigate Singapore's exhilarating transition into a global city, universally recognised as an enviable synthesis of technology, infrastructure, enterprise and manpower. It is a living plan that gives every individual and endeavour seamless access to intelligent technology—and with it—the capability to take charge. It is the new freedom to connect, innovate, personalise and create.

And although the language may hold echoes of a proximate future, what follows is a series of benchmark achievements and milestones, a clear role for centralized government, and the (re)setting of agendas.

Statistics tell one story. The IDA (2010a, 2010b), Singapore's government authority tracking information and communication technology adoptions, puts cell phone uptake at 138 percent and broadband at 157.9 percent. As of May 2010, there were clearly many more cell phones than people in Singapore, even though only 94 percent of the population owns a cell phone. In 2009, a total of 83 percent of Singapore's 1.02 million households had a computer at home, and all but 1 percent of those households were connected via some form of high-speed broadband (IDA 2010a, 2010b). Broadband connectivity extends well beyond the boundaries of the home. Both of Singapore's mobile phone carriers provide a large number of wireless hot spots, and consumers are connecting through these and the mobile data networks that the carriers also supply. In addition to these information and communication technologies, Singapore also has other sorts of ubiquitous technologies, including the world's first electronic road-payment system, a blanket closed-circuit-television network, fingerprint-only biometric banking, a computer-driven mass transit system, and an enviable smart-postal network.

It is not so difficult, then, to make the case that Singapore represents a kind of ubicomp environment. Indeed during the SARS epidemic in 2003, the government strategically employed the Internet and cell phone base as a way to distribute critical information as well as provide cyberalternatives to ongoing social practices, including the festivals around Ching Ming (also known as tomb sweeping day), which fell during the outbreak. The Singapore government helped create online shrines for the ancestors of Singapore's Chinese community, encouraging them to venerate at home rather than in Singapore's many cemeteries and, by doing so, hoping to cut the risk of the disease's transmission. In thinking about Singapore as a ubicomp environment, it is also important to think about what happens on and around the infrastructures built in accordance with IT2000 and iN2015 and those that just happened (like the uptake of high-speed broadband, total saturation of cell phones, and introduction of new 3G networks). What kinds of experiences, services, and applications transpire? What are people doing, or not doing, in an environment so full of sensing and smart technologies?

As we noted, almost everyone in Singapore owns a cell phone; it is unlikely that even those without phones are beyond reach of one. Phones are utterly pervasive and can be found interwoven into most aspects of daily life (Bell 2005, 2006b). The two major mobile service providers offer a wide range of mobile services from feng shui advice to the weather to home survelliance; they also offer various translation applications, which makes good sense in a country with four official languages (Chew and Kramer-Dahl 1999). The mobile nature of cell phones also creates new opportunities for location-based services. Everyone in Singapore jokes about the impossibility of hailing a taxi without a cell phone, for instance. Most public areas in Singapore now have a location-based numeric code displayed on wall signs allowing one to send a short message service (SMS) message to a taxi company, which then dispatches a cab to the location—simple and effective. Singaporeans send on average more than one billion messages a month and rely on their phones for everything from basic communication to more sophisticated forms of commerce, social interaction, and political engagement.

There are other ways in which Singapore's transportation infrastructure blends different sorts of technological interventions in both the public and private domains. Again, Singapore's size is an advantage. The vast majority of Singapore's residents use public transport in their daily lives. In fact, Singapore's mass rapid transit system serves more than one million passengers each day in the city's high-density travel corridors, and many more people use the bus lines to reach their final destinations. The train system utilizes smart card ticketing and advanced CCTV to provide seamless and safe transportation. Although only about 11 percent of the population owns a car, there is a significant infrastructure for taxis and traffic control in Singapore. In 1998, Singapore's Land Transport Authority implemented the world's first electronic road pricing system (Menon and Keong 1998; Menon 2000). Utilizing a dedicated shortwave radio communication system, unique in-vehicle identification units, smart cards, distributed data collection points, and a centralized data center, the electronic road pricing provides Singapore's drivers with variable pricing schemes for entry into the central business district during the week. The data are displayed on large LED boards at major intersections ringing the restricted zone; automobile drivers make decisions based on the pricing at these transition points and taxi drivers consult with price-sensitive passengers regarding the appropriate routes. The tolls are debited directly through a smart card system. Interestingly, while the system did not result in a boost to the public transportation system, within the first year there was a noticeable drop in the number of times any single vehicle entered the restricted zone on a given day (Menon 2000). Digitally enhanced vehicles, smart toll booths, and various forms of technological interventions in the mass transit system all point to an almost-seamless manifestation of ubicomp in the otherwise-prosaic domain of public transportation.

The development of an intelligent island/nation has not been without its challenges. For Singaporeans, this ubicomp environment—with its mobile handsets, pervasive Internet, and smart sensors—has raised questions about content, surveillance, and control. The regulation of ubicomp spaces is not often a well-documented subject in the relevant literature, save through the vectors of privacy and security. Yet the Singapore example seems to offer some interesting reworkings of the consequences of a wide-scale deployment and its subsequent accessibility. In late 1996, the Singaporean Ministry of Information and the Arts announced plans to filter all Internet use through government proxies to regulate access to political, religious, and pornographic content online. Not only were Internet service providers, cybercafes, and websites required to register with the Broadcast Authority but, adopting an approach similar to that used in China, the government also blocked access to at least one hundred sites that it deemed most problematic. Usenet groups were subject to scrutiny, intervention, and regulation as well (Ang and Nadarajan 1995). The head of the Singapore Broadcasting Authority was quoted at the time as saying that the regulation was in the service of national safety: "We don't want objectionable material to be easily available. . . . We just want to keep this part of the Internet within our immediate neighborhood clean." What is particularly interesting about these censorship moves is that many Singaporeans regard them as positive and appropriate government actions (ibid.; CRC 2003).

Singapore's censorship regime has been well documented (Heng and Devan 1992). Its seemingly collaborative nature, clearly articulated principles, and relative transparency make it worthy of closer scrutiny, however, as it has interesting implications in the ubicomp space. In 2003, Singapore authorities revisited these questions of the Internet's regulation and control in their decadal review of censorship in Singapore (CRC 2003). The committee defined censorship not as a confrontation between regulators and the regulated but rather asserted that "it is about collaboration to debate social issues constructively and to help educate members of a society to be more knowledgeable and sophisticated" (ibid., section 9.4). In his letter to the chair of the Censorship Review Committee, the minister of information, communication, and the arts at the time wrote:

> Your committee has succeeded in keeping the report relevant against the backdrop of our social evolution and changing global landscape. While understanding the need to fan the creative flames of the new generation and to accommodate the diversity of views, you were sensitive not to weaken the "glue" that bonds our society—our core values, our identity, our shared memories, our religious and racial harmony. Your committee also sees censorship as a shared responsibility, thus your approach of encouraging participation comprising regulators, industry players, community and artists in the process. . . . Our common challenge is to achieve a balance where adults can have wider access, whilst our young are provided with a conducive environment to develop morally and socially, without compromising the development of creativity and social capital. These objectives might be difficult but certainly not incompatible. (ibid., ii–iii)

Here technology is understood as always and already operating within a cultural context, and a complicated one at that—the religious, cultural, and racial heterogeneity of Singaporean society is a constant feature of negotiations around the appropriate regulation of technology environments. This construction of censorship happened in the larger context of Singapore's social/cultural regulation of the Internet, in particular when the National Internet Advisory Committee (NIAC 2004, 1) continued its charter to promote "the safe and positive use of the internet." In 2002, the committee reported ongoing concerns about Internet accessibility via mobile devices, especially when they negated the value of parental supervision. While it is possible for parents to put the personal computer in the communal room at home to supervise their children's online activities, such a safety measure is greatly diminished with a mobile Internet (ibid., 4). This continuing concern with children's safety in particular prompted the NIAC to launch a new "cyberwellness" initiative. "In the cyber wellness vision, users should not only embrace the Internet in their lives, they should also adopt an attitude of using the Internet for inspiring others and contributing to the community" (ibid., 7). The committee went on to suggest that this vision of cyberwellness also incorporates a sense of personal responsibility—to not spam, to not spread misleading information, and to verify the accuracy of information gleaned online before acting on it (ibid., 7). This language and its intent is remarkable. Here, the ubiquity of the Internet is seen to support collective and community practices, and individuals are not just satisfying their own desires and needs but instead operating within a larger cultural framework. Imagining ubicomp as a collective practice as opposed to a set of discrete individual actions is an important reframing of that technological vision.

Everyday Ubicomp: Korea

If Singapore is an example of collective uses as well as computational device and sensor ubiquity, then Korea is an instance of infrastructural ubiquity and of public/private sector cooperation to achieve it. As we reviewed earlier in this chapter, the Republic of Korea has an explicit government mandate to create a ubicomp society. With its population of slightly more than 49 million, Korea routinely ranks as one of the most connected countries in the world: 84.4 percent of Korea's 17.3 million households have computers in them, and almost all of these computers are connected through some form of high-speed broadband. The average user spends about fifteen hours per week using that computing, with eleven of those hours online. Just over 75 percent of the population reports using the Internet regularly; a staggering 96 percent of Koreans ages six to twenty report using the Internet regularly—for everything from email to gaming, social networking, blogging, religion, health and data management, and education (NIDA 2004, 2007). Furthermore, according to the National Internet Development Agency of Korea, 83 percent of Koreans have their own email addresses—including 85 percent of those under twenty, 91 percent of those in their twenties, and a remarkable 67 percent of those over sixty—and just under a third of all Koreans who use the Internet report having more than one email address. In terms of more recent digital phenomenon, 40 percent of all online Koreans report managing some kind of blog, with female users outstripping male ones—including more than 45 percent of those in their teens, more than 68 percent of those in their twenties, 40 percent in their thirties, and a full 9 percent of those over the age of sixty (ibid., 2007). It is safe to say that the Internet is a ubiquitous technology in Korea.

Korea is also frequently touted as the leading broadband market in the world. Korean broadband is characterized by high-speed downloading and a comparable backhaul, creating a digital environment that is both consumptive and participatory. The remarkable level of connectivity is in no small part facilitated by the nature of Korea's urban landscape. More than 81 percent of Koreans live in urban areas—indeed slightly more than 20 percent of the country's population lives in Seoul alone—and most of them live in high-rise, multifamily, high-density dwellings. These tall, heavily populated buildings (evoked so beautifully on the Ubiquitous Dream Hall website as treelike living node points) create a last-mile boon, not available in U.S. urban sprawl: you only need extend the wire to a building, and plug the whole building in, rather than wiring house by house. As a result of readily available and relatively cheap high-speed data connections (in both directions), Koreans enjoy a wide range of Internet usages at home, including watching previously shown television programs, robust local-content virtual worlds and social network sites, health and medical services, home banking, stock trading, and even church services streamed to their home computers. Finally, the Korean government holds Internet service providers financially responsible for spam and viruses originating on their networks, helping to drive a different kind of Internet

In addition to this robust home broadband ecosystem, Korea also happens to be one of the fastest-growing markets for PC Bangs or cyberarcades—gaming parlors with between twenty to two hundred machines, designed to support online gaming. These arcades have flourished even as Korea's home-personal-computer uptake has grown. This seemingly topsy-turvy reality makes sense when you know that Korean homes are considered to be extremely private domains, often closed even to one's closest friends, and that socializing, especially when it comes to gaming, has nearly always had a space in the public domain and is in fact actively sought out that way. While 90 percent of Korea's online population goes online at home, 25 percent report regularly going online in cybercafes too. Even those people who have computers at home are also frequently using computational devices away from those homes. Further eschewing some of our expectations of technology-usage patterns, of the sixteen million or so Koreans who make use of some form of mobile Internet access, only 30 percent do so on the move; a full 40 percent report using their mobile devices to access the Internet while at home (NIDA 2004)—and this in the days before iPhones and other smart mobile devices that made Internet access seamless. Interestingly, women are slightly outpacing men in their consumption of mobile Internet experiences—38 to 34 percent, respectively. Less surprising, Koreans using mobile Internet access skew young: 83 percent of those under twenty report using it at some time, and this includes everything from surfing the web to engaging in a range of e-commerce and m-commerce, where the phone is the dominant mode of payment for online content in any form.

In 2004, against this backdrop of ubiquitous connectivity, devices, and content, and also a booming middle class (Lett 1998; Nelson 2000), Daeje Chin, the minister of information and communication and a former senior executive at Samsung, launched IT839—"The Road to $20,000 GDP/Capita"—a remarkable technology plan that proposed to transform Korea into a ubiquitous society within five years (MIC 2004). The ongoing global financial crisis has impacted this plan, delaying some of its impact, but it has not changed the focus on the centrality of technological ubiquity in the Korean governmental imagination. "U-Korea," as it was quickly dubbed in the press, was a bold plan that relied on the close ties that already exist between Korea's government and many large private sector companies (C. S. Kim 1986; E. M. Kim1997; Koo 1993). Much of the plan's tenets are spelled out on the Korean Home Network Industries Association's website, whose graphics bear an interesting resemblance to those on the sites for the Ministry of Information and Communication.

The plan, revisited and revised in 2005 (MIC 2005), takes its name from eight services (Wireless Broadband, Digital Multimedia Broadcasting, Home Network, Telematics, RFID, W-CDMA, Terrestrial Digital TC, and VOIP), three infrastructures (Broadcast Convergence Network, Ubiquitous Sensor Network, and Next-Generation Internet Protocol), and nine equipment fields (Next-Generation Mobile Communications, Digital TV, Home Network, IT System-on-Chip, Wearable PC, Embedded SW, Digital Contents, Telematics, and Intelligent Service Robot). It is designed to boost

Korea's production, employment, and exports, ultimately raising the entire country's gross domestic product as well as those of all households. In the press, Chin was quoted as saying that U-Korea meant "a society where all people can enjoy the benefits of state-of-the-art IT at anywhere and anytime"; more recently, "happiness" has made a significant appearance in the rhetoric. It appears to be well understood that the ubiquity to which Chin refers is a socially experienced one that happens to Korean society, not just Korean individuals. Indeed, even the metric of IT839's success is at a household level—per capita household income will rise, as will the quality of life for all Koreans. The vision of a technology future, here explicitly called out as ubicomp, is again manifested at a collective cultural or societal level.

Designing Ubicomp

For those who approach these topics from outside the core domains of ubicomp, HCI, or allied disciplines, it is important to note a fundamental consideration that is central to the ubicomp reseach agenda. Whereas in many areas it makes sense to talk of a separation between research and design, in ubicomp and related areas of experimental computer science, the two practices are much more tightly coupled. Those activities that we might most think of as research activities—such as experimentation, empirical investigation, and model and theory building—are in ubicomp frequently conducted in concert with design practices—identifying needs, giving form to new objects and responses to meet them, and crafting objects and systems. Sometimes design is the means for research activities to be carried out; that is, the process of design is the means by which ideas are explored and tested. At other times, and perhaps more commonly, design is seen as the natural end point of research activity; that is, the kinds of understandings derived from research are evaluated primarily in terms of the contributions they might make to the production of new artifacts. To those inside the discipline, this role for research and theorizing is so taken for granted that it is hard to contemplate or recognize any alternative (Dourish 2006a).

This attitude is not universal within computer science, but it is certainly prevalent in engineering and experimentally oriented areas of systems research such as ubicomp, and it is also the culture of some of the places where ubicomp arose. The scientists at PARC certainly lived by the motto "Build what you use, use what you build," expressing their commitment to make their research activities take actual form so that those experimental systems could be used in everyday settings to understand their implications. Similarly, Philip Agre (1997) documents the same culture at work in the Artificial Intelligence Laboratory at MIT (now the Computer Science and Artificial Intelligence Laboratory), where he conducted his doctoral research, and discusses the ways in which the engineering discipline as he learned it systematically undermined his early attempts to understand relevant writings from the social sciences and humanities. He notes that to an engineer, for instance, any approach to theoretical and analytic investigation is generally read as specification—that is, as an attempt to uncover and specify the kinds of mechanisms that can be implemented in computer systems—an approach often at odds with the fundamental topic.

It is not simply that design activities play a critical role in ubicomp research but rather that design, for many, is the end goal toward which ubicomp research is directed. As such, it is useful to distinguish between ubicomp as a research agenda—a set of goals and objectives for research endeavor—and the particular visions of ubicomp embedded in specific designs. Ubicomp's linked concerns with research and design sometimes obscure this, but there is value in maintaining at least an analytic separation between the two when examining the rhetorical positioning of both traditional research efforts and design practice. The nature of the hybrid research/design model means that we often can see design efforts as attempts not to concretize the outcomes of research but instead to justify, promote, or initiate them; likewise, it can give rise to systematic misreadings of analytic statements whose focus is not design practice overtly—such as ours here. An account of ubicomp as messy, for example, is not something that we propose to drive design; rather, we see it is a critical contextualizing remark for helping to understand the limits and conditions on design along with the nature of the settings into which designed artifacts and infrastructures are placed.

Toward a Ubicomp of the Present

There are a range of differences between the settings we have been describing in this chapter and those that more conventionally appear in ubicomp research, and these may explain the absence of these sorts of accounts as parts of the ubicomp vision. Perhaps it is that in Singapore computational technology is both embedded in daily life (e.g., closed-circuit television or sensors) and carried around (e.g., cell phones), disappearing into the fabric of everyday life. Perhaps it is that Korea's infrastructural development relied not on the free market but rather on a strong relationship between the public and private sectors. What this suggests, then, is an alternative domain of ubicomp research—a ubicomp not of the future but instead of the present. William Gibson (1999) famously quipped that "the future is already here; it's just not very evenly distributed." We take each component of this aphorism as a piece of an alternative research agenda for ubicomp.

The future is already here. The technological trends that Weiser insightfully extrapolated have resulted in radical transformations and reconfigurations of the role of computation in everyday life, just as he anticipated. Weiser foresaw a world in which computation would be embedded into our daily worlds—not just physically embedded, but also socially and procedurally embedded, becoming part and parcel of how we act. It has not, perhaps, taken the form that he expected, although personal digital assistants, cell phones, large-scale displays, and digital cameras do bear family resemblances to the devices that Weiser imagined would come to populate our society. Nevertheless, the fact that the details are different should not blind us to the remarkable accuracy of Weiser's vision. Computation is embedded into the technology and practice of everyday life; we continually use computational devices without thinking of them as computational in any way. The desktop computer has not been displaced but instead augmented.

Interestingly, though, while the technological form of ubicomp differs only in its details from the model that Weiser anticipated, it is perhaps the *use* of ubicomp that would have surprised him. One notable aspect of Weiser's article is that as he lays out a vision for a radically different form of computational experience, the settings into which those devices are to be deployed and the activities that they are used to support remain largely unexamined. Weiser's ubicomp technology is still used primarily in workplaces, relies on large fixed infrastructure investments by commercial entities, and is directed toward the needs of corporate efficiency. Other manifestations, like those sketched earlier in this chapter, certainly suggest the possibilities of domestic and other nonwork settings.

From the perspective of a ubicomp of the present, we can note that Weiser was entirely correct in one regard: the purposes to which people would put computational devices are not radically new ones but rather reflect existing social and cultural practices. However, this did not necessarily take the form that had been anticipated. Computational technologies are embedded in social structures and cultural scripts of many sorts; ubicomp technologies prove also to be sites of social engagement, generational conflict, domestic regulation, religious practice, state surveillance, civic protest, romantic encounters, office politics, artistic expression, and more.

What this suggests is that we need a deeper understanding of how social and cultural practice is carried out in and around emerging information technologies. If ubicomp is already here, then we need to pay considerably more attention to just what it is being used to do and its effects. Interestingly, while considerations of the social and cultural elements in ubicomp's agenda have traditionally been thought of in terms of social impacts, our focus in this book is more on technology as a site of social and cultural production—that is, as an aspect of how social and cultural work are done as opposed to something that will inevitably transform social practice. It may be quite the other way around.

The idea that the future is not evenly distributed has traditionally been taken to note differential access to technologies, the digital divide, and in particular, the concentration of advanced technological infrastructures within the research laboratories of universities and corporations. Here, we read it in three ways: first, as noting the ways in which power relations are embedded in access to infrastructure; second, as pointing toward the quite different patterns of technology adoption and use in different cultural settings; and third, as signaling a primary concern with how inherently messy and uneven infrastructures are encountered and navigated.

For one, drawing on analyses of "time-space compression" such as that of David Harvey (1989), Doreen Massey (1993) coined the term "power-geometries" to refer to the ways that spatial arrangements (e.g., the locations of homes and their proximity to both amenities and sources of noise and pollution) as well as patterns of access and mobility (e.g., in the competition for resources between different forms of public and private transportation) reflect arrangements of power and control. These power geometries also affect the relationships between spaces and the means by which those relations are brought about. For instance, reflecting on the area of London where she lives, Massey comments: "It is (or ought to be) impossible even to begin thinking about Kilburn High Road without bringing into play half the world and a considerable amount of British imperialist history" (ibid., 65). Similarly, when we think about ubicomp technologies and infrastructures, the patterns by which they are introduced into existing spaces, and the needs around which they are designed and for which they are harnessed, we are immediately presented with the need to acknowledge the ways in which technologies both exploit and reproduce a range of power concentrations and relationships.

Next, when we see ubicomp as a feature of the present rather than of the future, then we are forced to contend with it as an inherently heterogeneous phenomenon. Standardization and consistency can be imagined in the future, but a technology of the present is one that operates in a thoroughly heterogeneous environment. This applies at all levels. Consider, for example, the radically different forms of governmental processes at work in a comparison between U.S. and Singaporean approaches to Internet information provision (or even, for that matter, between U.S. and European approaches to mobile telephony regulation). At the same time, on a much smaller scale, the interplay of standards for telecommunications render an apparently "invisible" infrastructure highly visible in terms of the range of concerns to which its users must be oriented. A technology of the present is inherently one that is deployed and operated in a fragmented world.

Third, and relatedly, focusing on ubicomp in the present as an inherently heterogeneous phenomenon suggests that a significant aspect of its research agenda should be the ways in which heterogeneity is currently manifested and managed. We have discussed ubicomp as inherently messy. So how do people both manage and even exploit this messiness in current interactions? From an ethnographic perspective, this might take the form of Star's (1999) ethnography of infrastructure, using the multiple manifestations of infrastructure as an analytic lens as in, say, Star and Karen Ruhleder's (1994) exploration of the infrastructure of scientific collaboration. From a technological perspective, it might take the form of the seamful design examined by Matthew Chalmers and Areti Galani (2004), in which the variability of infrastructure and access become a resource for active engagement.

We seek to engage with these themes throughout this book. Our attempt is to understand ubicomp in all its delightful messiness and to see how current ubicomp—in both the everyday world and the contemporary scientific imagination—is socially and culturally constituted. In this chapter, we outlined the high-level program and provided some broad examples. We turn next to methodological questions to refine our topic and the ways to approach it, then in part II follow our broad themes through a range of more specific topical sites.

3 Making Room for the Social and Cultural

In the previous chapter, we discussed the history and status of ubiquitous computing (ubicomp) as a research and development endeavor, and also what opportunities and implications are involved in attempting to approach it from an interdisciplinary perspective. Clearly, we do not believe that the story ends when the technology leaves the laboratory—nor does it start within the bounds of the laboratory in the first place. We are concerned with its embedding within social and cultural settings, and what kinds of implications we should draw from the nature of that embedding.

Most of the book—and in particular part II—is devoted to just that analysis. Before we begin that, though, it is worth spending some time exploring what we have in mind when we talk about offering social and cultural accounts. We are obviously in the broad realm of the social sciences here, but just what sort of perspective is offered? Are social accounts and cultural accounts different? What might each look like? What's cultural, anyway?

It is almost impossible to give universally applicable answers to these questions because to the extent that they are terms of art in disciplinary contexts, the "social" and "cultural" both name bodies of work that have complicated histories, and ones that have played out quite differently in different places. Throughout this book, we look to anthropology, sociology, and ethnography as the significant hubs of the social sciences.

What we want to do in this chapter, by way of preparation, is to outline the take that we will offer here on the social and cultural. There are some important differences, both epistemological and methodological, between our approach and that which is often at work in ubicomp. These differences concern the nature of the relationship between disciplinary interests that play out in ubicomp research. To say that we are interested in the social and cultural aspects of ubicomp, then, is not to say that we are interested in what happens when ubicomp moves into the world; rather, we are interested in how it is already a sociocultural object, both in its artifacts and practices.

Invoking the term sociocultural is a hedge, but it is one that reflects a complicated historical entwining of various disciplinary interests. In the United States, for example, anthropology has traditionally been defined as a discipline with four major fields or subdisciplines: linguistic anthropology, physical anthropology, archaeology, and cultural anthropology. At the same time, what is generally thought of as cultural anthropology in the United States usually falls in the scope of social anthropology in, say, the United Kingdom. This difference in nomenclature to some extent reflects the different contexts in which the discipline of anthropology emerged in those two countries. On the one hand, the social might be thought of as primarily the domain of sociology while the cultural is largely that of anthropology; much important and creative work, on the other hand, has taken place in institutions where these two disciplinary perspectives interact (such as the University of Chicago), while the emergence in the second half of the twentieth century of cultural studies—itself something of an umbrella for a variety of academic perspectives—has troubled anthropology's claim to define culture as its own particular object of study.

As a starting point, let's broadly, provisionally, and problematically suggest that the academic study of the social focuses on the scales and structures of social life—the patterns and functioning of social institutions, the interrelationships between social structures and their evolution, and questions of the stratification, segregation, and distribution of social resources, including money, power, influence, and authority. Equally broadly, provisionally, and problematically, let's suggest that the academic study of the cultural concentrates on the social as a symbolic form and on the ways in which people find these things meaningful—the conditions under which specific kinds of collective value and significance attend to the ritual practices of everyday life, the world as we encounter it, and the ways in which we interact.

This rough-and-ready distinction might help us to understand a little of the nomenclature on either side of the Atlantic. British anthropology has frequently concerned itself with social organization, which cannot be entirely separated from the British colonial enterprise and the way in which anthropologists were recruited to help understand the social structures through which colonial governance might operate; at the same time, in the United States, anthropology arose not least in the context of the dwindling numbers of Native Americans, often with a focus on language and meaning (Stocking 1982, 1991, 1992; Trautmann 1988). So some affinities with the social and with the cultural are perhaps rooted in these disciplinary and institutional histories. Sometimes confusingly, British sociology looks surprisingly like cultural anthropology in the United States, though it clearly has antecedents in profoundly different intellectual histories and questions and usually finds itself in dialogue with questions of public policy, social good, and government regulation.

Across the Atlantic and between these different social science hubs, it is immediately apparent that the social and cultural cannot be separated. The structures of social life themselves have meaning, and meaning is a structural phenomenon, so any principled and absolute separation is meaningless. Similarly, anthropologists have spent much time investigating social structure (perhaps most characteristically, patterns of kinship), while sociologists from perspectives such as symbolic interactionism (Bruner 1969) have been deeply concerned with the production of meaning and signification. So this depiction is too rough-and-ready, accounting perhaps more for the "center of mass" of each view, while admitting significant overlap in methods, topics, and concerns.

Fitting in Ethnography

Ethnography has its roots within the discipline of anthropology. As we flagged a moment ago, anthropology itself has its origins in the Western expansionism of the nineteenth century. In North America, this took the form of the "salvage anthropology" of Franz Boas and his colleagues, documenting the rapidly shrinking cultures of Native Americans; it was associated elsewhere with colonial encounters between Europeans and the peoples of South America, Africa, the South Pacific, and beyond (Stocking 1982, 1985, 1991, 1992, 1995). Ethnography itself arose in the early twentieth century, spearheaded not least by Bronislaw Malinowski (1922) in his work on the Trobriand Islands. The emergence of ethnographic investigation marked a major transition in the practice of anthropology, emphasizing as it did the need to understand the member's point of view. Where anthropology had previously documented what members of other cultures *did*, ethnography argued that through daily participation in everyday life, one could come to understand what members of those cultures *experienced* through their actions. In contrast to surveys and interviews, ethnography advocated long-term, immersive fieldwork combining observation with participation. Ethnographic field research became the hallmark of anthropological inquiry in the following decades, and generating holistic, culturally specific monographs became a stock in trade and academic tenure requirement (Knauft 1996; Bohannan and Glazer 1973).

In the second half of the twentieth century, and especially over the last quarter century, anthropological traditions in the United States, the United Kingdom, and other key world sites have refined and reworked a set of critical epistemological concerns around questions of reflexivity, voice, stance, and standpoint. It is worth noting that in this same period, North American anthropology developed a complicated relationship to interdisciplinary collaborations—in particular across the social science/science divide as well as the industry/academy one. Deep mistrust of industry along with the attendant characterizations of "applied work" (in a non–nongovernmental organization space) arose out of 1950s–1960s' relationships between anthropologists and the U.S. defense and state departments through such collaborations as Project Camelot (Horowitz 1967; Madian and Oppenheim 1969) and Ruth Benedict's early *The Chrysanthemum and the Sword* (1946).

Ethnography has also migrated into other social sciences, albeit with different flavors and emphases. The work of the Chicago School sociologists (Robert Park, Everett Hughes, Herbert Blumer, Anselm Strauss, Howard Becker, and more) was influential in bringing ethnographic methods into sociological research. Much of the emblematic work of the Chicago School had two characteristics. First, it turned its ethnographic attention toward aspects of U.S. urban life, applying anthropological methods to issues in locales quite different from those in which Malinowski had developed his methods. Second, it retained its sense of ethnographic distance between subject and object through inquiry into subcultures and "outsider" groups (such as tramps, prostitutes, and gamblers) and odd locations (such as public toilets and mental institutions). The uptake of ethnographic methods to study settings and people "closer to home" is arguably a critical step toward the use of these methods to study technology users.

This is a partial view, and in the course of what is to come, we will have more to say about both the history of ethnography within anthropology and its disciplinary considerations, and about the ways in which ethnography has been appropriated within ubicomp.[1] This history, though, sets a context for understanding how ethnography has been adopted in an encounter between social science and technological considerations in ubicomp research and for reflecting on some of the consequences.

The Social

While the primary sources of energy and investment in research in information and communication technology focus on engineering concerns, the social study of information technology has been a minority but consequential interest for several decades. This work has arisen in different domains and with different orientations.

One approach has concerned the consequences of technological advancement on the social settings in which technology is deployed—the workplace and factory floor, schools, domestic environments, and so on. Much of this work evinces an interest in the "social impacts" of information and communication technologies—the transformations that are wrought in organizational structure and practice, patterns of communication, and so forth, in the wake of the arrival of various forms of technology. Classic accounts include examinations of the ways in which digital technologies affect power relationships between technicians and scientists in medical imaging (Barley 1986) or studies of the ways in which electronic mail revises communication patterns in organizations (Sproull and Kiesler 1991).

The social impacts perspective reaches beyond simply the ways in which applications are adopted; it encompasses, too, the underlying constraints that technologies introduce. Take the Internet as an example, and in particular the technical protocols that govern how data traffic flows from host to host and network segment to network segment. A social impacts account of the Internet protocols would perhaps give us some insight into the ways in which social phenomena organize themselves around the features of this protocol. For instance, certain features of the mechanisms for the allocation of network addresses ("IP addresses") make it easier for large organizations to host content than for individuals to do so because most strategies for hosting content require the allocation of

---

1. In this same period, anthropologists initiated more earnest ethnographic inquiries into sites where ubicomp researchers worked and their research agendas were being set. Science and technology studies started to make ethnographic inquiries around the edges of ubicomp with works by Julian Orr (1996), Stefan Helmreich (1998), and perhaps even Sharon Traweek (1988). Within ubicomp, given its historical connections to domains such as HCI, one significant strand of ethnographic work has been ethnomethodological in orientation—often quite dogmatically so (e.g., Crabtree et al. 2009)—and so largely disconnected from the questions we address here.

fixed Internet addresses to information resources, and static addresses are allocated in large blocks that are useful primarily to organizations and larger institutions. Individual subscribers are more likely to use smaller numbers of addresses (having just one or two computers) and so are more likely to be allocated addresses dynamically from a pool reserved by a service provider. This is a more efficient way to populate the address space, but it means that in the absence of other compensatory mechanisms it is more difficult for individual subscribers to host servers and provide information at known locations in the system.

A naive orientation toward social impacts, though, frames the relationship between the social and technical too narrowly. We should be wary of approaches that take information technology as a given and society as a passive recipient (or impact site) of technology that arises independently and inevitably. Most especially, it is important to avoid the perils of what Rob Kling and his colleagues (2000) refer to as the "layer cake" model of the relationship between technology and society, in which social phenomena play out independently of and layered on top of a predefined technological substrate.

We can return to the example of Internet addressing to illustrate this. When we study the social impacts, we take the technology as it exists and ask, "What are the consequences?" But what this misses is the social context within which the technology was designed and produced in the first place. What of the context in which the Internet protocols were developed? The Internet protocols were developed in the military and academic research community of the 1970s and early 1980s. The driving motivation for the Internet (then ARPANET) was access to large-scale computer resources at a national level, with a military interest in the resilience of command-and-control infrastructures to attack. As a result, a series of assumptions worked their way into the protocol design: that computers were rare and expensive devices owned by large institutions with the resources to employ people to manage them, that the network changed relatively slowly and computer systems did not move around, that people also would be found at relatively stable and fixed points in the network, and that the topology of the network was generally stable. The protocols that emerged were ones that reflected these assumptions. They provided for addressing and routing in relatively stable environments, in which the primary problem was to route information between organizations rather than among individuals.

Here, we have a different perspective—one that focuses less on the ways in which technology may have impacts on society and more on the ways in which society, both its structures and assumptions, has an impact on technology. This alternative point of view—the "social shaping" of technology—recognizes that technologies in general (and information technologies in particular, in our case) are the outcomes of social action. They are generated by people operating in social contexts (of all different sorts) and at particular historical moments, all of which shape the imagination of what needs technology might meet and in what settings it might be employed.

The Cultural

As suggested earlier, to the extent that it makes sense to distinguish between the social and cultural (and that is only a limited extent), the concept of culture that we discuss here is largely a semiotic one, concerned with the ways in which we find meaning in the world. Clifford Geertz (1973, 5) provided perhaps one of the most classic descriptions of this symbolic account of culture:

> The concept of culture I espouse . . . is essentially a semiotic one. Believing, with Max Weber, that man is an animal suspended in webs of significance he himself has spun, I take culture to be those webs, and the analysis of it to be therefore not an experimental science in search of law but an interpretative one in search of meaning.

The concept of culture, what it means, how it is used, and in what contexts it might be deployed is by no means cut-and-dried, and even within anthropology it has been the topic of repeated—and not entirely conclusive—debate (e.g., Kluckhohn and Kelly 1945; Kroeber and Kluckhohn 1952; Keesing 1974). Some of this discussion is essentially recapitulated in terms of how the concept has been used by technologists. As designers of information technology have increasingly come to recognize its significance to their concerns, questions of culture have manifested themselves in the technological research literature (e.g., Kayan, Fussell, and Setlock 2006; Marcus 2001; Massey et al. 2001; Setlock, Fussell, and Neuwirth 2001). Yet different approaches manifest different interpretations of culture as an analytic category. We will make a distinction between two here: a taxonomic account, and a generative account.

The taxonomic account of culture is one in which culture is deployed as a means to categorize people, groups, behaviors, attitudes, and beliefs (Boas 1940). This taxonomic account lies at the heart of attempts to identify the consequences of "cultural difference" as a consideration in the uptake and use of technologies, and hence to design for different cultural settings. What is important in this account is the ways in which cultural settings can be enumerated, differentiated, and then analyzed in terms of their properties. So in this model, one might distinguish between British and U.S. culture, northern and southern European culture, Asian and Latin, and so forth. One might then discuss how values such as thriftiness, modesty, generosity, friendliness, independence, respectfulness, diligence, stoicism, and so on, are differently valued, ordered, and expressed in these different cultural settings. This approach still finds currency in the work of people such as Gerd Hofstede (2001), whose model of five dimensions of culture is routinely recommended as an analytic tool for making sense of, and thus better managing, cross-cultural organizations and institutions.

This account of culture is one that we might associate with anthropology, particularly that of the early twentieth century, during which time, under the broad label of "area studies," anthropologists and others worked to catalog systematically the breadth of human cultural experience as well as to understand the parameters of cultural diversity. Researchers became area specialists, expected to be able about to talk about cultural patterns and practices in their designated area to a high degree of specificity. Over time, however, and especially given the rise of semiotic perspectives on culture such as that expressed by Geertz, this approach began to founder on a number of key problems.

The first problem is, Where should the boundaries between cultural regions be drawn, and how sharply? Do we follow the boundaries of nation-states, and if so, what do we do about the fact that these are often relatively recent inventions (and still frequently in flux)? Do we instead draw on patterns of religious observance or language use as ways to distinguish between cultural groups? What is more, even when those boundaries are drawn, how do we handle the inevitable traffic in culture across boundaries and divisions, however they are drawn? Borderlands of all kinds, even those rigorously policed, are inevitably sites of hybrid cultural practice. If one were to distinguish between Thai and Burmese culture, Russian and Chinese culture, or Eastern European and Scandinavian culture, what can be said about those places where those come together? The taxonomic account of culture is central to a search for essences, and yet to distill cultural practice down to a mix of such essences is a deeply problematic task.

The second problem is, How is this to be understood in an increasingly visibly interconnected world? Today we might think in terms of cheap air travel and the reach of global media, but trade and exploration have long put peoples in touch with each other, resulting in complex cultural hybrids of all sorts—from Aboriginal Australians whose language includes loanwords in Arabic (through their contact with Afghanis who came to Australia as camel drivers [Ganter, Martinez, and Lee 2006]), to the linguistic and cultural remnants of Viking exploration throughout the North Atlantic region. Questions of cultural "purity" and "authenticity" are problematic here, and deeply perspectival too; Americans who lament the "cultural imperialism" of McDonald's opening in Beijing speak at the same time of the cosmopolitanism associated with Chinese restaurants in Georgia (Lu and Fine 1995), while newly minted middle-class Chinese twenty-somethings take their dates to those selfsame McDonald's in Beijing and Shanghai, attracted by all that auspicious red and gold coloring (Robison and Goodman 1996).

The third problem is, When we start to think of culture as a semiotic system, we are forced to recognize that each of us pulls from many cultural systems in order to understand our worlds. Yes, perhaps our national origin is associated with some cultural values that our daily lives express; but so are our religious affiliations, professional training, gender, socioeconomic status, and biographical experiences. The university professor, physician, and rabbi are each enmeshed in Geertz's "webs of significance" that give their worlds meaning, much in the same ways as the Scot, Australian, or Hmong.

Problems such as these lead us to a different view of what culture might be as an analytic category. This alternative to the taxonomic view is one we call the generative view, in which we think of culture as generative or productive of everyday experience—that is, we encounter the world through cultural lenses, which bring it into focus in particular ways while also rendering it meaningful and accountable to us. These lenses frame what we see, and how we see and understand it.

As an example, take a generative cultural reading of personhood. From a Western perspective, personhood is a given; it is a natural truth. It is certainly something we are born with; we have distinct and clear individual identities. Within other cultural frames of reference, though, personhood arises out of a complicated negotiation of ritual and profane activities, and indeed sometimes the notion of a strongly individuated personhood is neither a desirable end state nor a recognized trajectory. Writing about the Hagan, who make their homes in Papau New Guinea, Marilyn Strathern (1988) suggests that personhood in the Melanesian context exists only in and through relationships with others. One is not a person first, as might be true in the Enlightenment tradition; instead, one is a person only by virtue of one's positioning in a nexus of social relatedness. For the Sambia, who also make their homes in the eastern highlands of Papau New Guinea, achieving personhood is not just relational but also involves the actions of many others. For young Sambia men, achieving recognizable personhood involves all of the village's males making real and ritual sacrifices—personhood is not achieved solely through one's own actions; others are also implicated (Herdt 1994).

The generative account, as we have laid it out here, suggests that when we think about culture, we might want to concentrate not so much on a set of values and properties but more on how culture instead operates and is enacted in everyday practice. Our focus is not on culture as an object but on cultural practice as both active and performative. This active rather than passive approach is, similarly, an approach to culture that has been important in certain aspects of cultural studies that are particularly relevant to our concerns in this book.

The Cultural in Cultural Studies

We turn to cultural studies in order to carry these ideas forward in two ways: first, in terms of the continuing elaboration of what the topic and investigation of culture might be; and second, to further examine the idea of culture as an active process of meaning making.

As a discipline, cultural studies traces its origins to the work of the Frankfurt School, a group of influential neo-Marxist thinkers at the Institute of Social Research at the University of Frankfurt (Martin 1996; Wiggershaus 1995). The Frankfurt School's interests were by no means restricted to cultural matters, but they were a significant element of its reassessment of the foundations of Marxist thought for the mid-twentieth century. In particular, one component of classical Marxism that the Frankfurt school rejected was the distinction between the base and superstructure in Marxist theory along with its consequences.

For Karl Marx (1859/1977), this distinction was an account of the primal role of economic and material conditions in politics and society. The base, in Marx's thinking, is the level of economic and material activity, out of which flow all the consequences for the superstructure, which includes political and legal frameworks, cultural practice, and ideology. The relationship between the base and superstructure in classical Marxism expresses the role of economic and material productive forces as the fundamental conditions for society. Writers such as Max Horkheimer, Theodor W. Adorno, and Walter Benjamin, however, revised this relationship in order to place greater emphasis on the political role of cultural products, especially the emerging mass media. They were specifically concerned with the ways in which mass media might be used to promote hegemonic values and dismantle the conditions that lead to the emergence of a revolutionary political consciousness. The Frankfurt School turned its attention to the study of mass media, but it did so in a manner that was largely concerned with the critique of mass media as a tool of political repression.

Among the various sites that gave rise to contemporary cultural studies was an influential group of thinkers at the Center for Contemporary Cultural Studies (1964–2002) at the University of Birmingham in the United Kingdom, among them Raymond Williams, Stuart Hall, and Dick Hebdige (Turner 2002; Webster 2004). Like Adorno and his colleagues in Frankfurt, the Birmingham group was concerned with the study of culture as manifested in mass media, and again like Adorno and his colleagues, it studied this in a neo-Marxist tradition, albeit one that drew more heavily on Antonio Gramsci's examination of hegemony and ideology. Where the Birmingham School's work diverged significantly from that of its predecessors, though, was in the active role that it ascribed to consumers of mass media. Where Adorno and others had seen people as the passive recipients of the messages and values encoded in the products of the culture industry, the Birmingham School and its successors (particularly in the United States) argued that audiences are not entirely passive but rather actively appropriate and reinterpret the media products that they consume.

This happens on multiple levels and takes multiple forms. Certainly, we can see this sort of cultural appropriation by audiences at work in studies of fan communities, which frequently produce derivative works (including fan fiction and videos) that reinterpret, often radically, the motives, values, and meanings of the characters and plots on offer (Jenkins 1992).

More subtly, we may see audiences as "authors" of the emergent meaning of cultural products in those cases where minor characters are celebrated and turned into the central figures—to which the producers of media must then respond. (For instance, while Inspector Clouseau turned out to be the main character of the Pink Panther movies, the original producers imagined jewel thief Sir Charles Lytton to be the protagonist. On television, consider the case of "breakout characters" such as Arthur "The Fonz" Fonzarelli on *Happy Days*, Steve Urkel on *Family Matters*, and Angel on *Buffy the Vampire Slayer*. The first two began as minor characters but were elevated to central narrative positions on the basis of audience response, and the last moved away to Los Angeles as the central character of his eponymous spin-off, which ran for an additional season after the final closure of Sunnydale's Hellmouth.)

Less instrumentally but more pervasively still, this perspective contends, audiences continually juxtapose, appropriate, and interpret the images, situations, and characters of mass media products and play an active role in the production of the cultural meaning that these media products hold. The relationship between production and consumption is troubled by this relationship, but even more so is the division between active and passive roles in media production.

These considerations are relevant in the world of digital media, especially given the interests in "user-generated content" associated with Web 2.0 and related topics at the time of this writing. Many commentators have asserted that technologies such as blogs and sites such as the video-sharing service YouTube represent the progression of this trend, celebrating end users as content creators even more explicitly than in the terms of the cultural studies of mass media (Benkler 2006; Shirky 2008). That said, the Frankfurt School critiques remain important for any account of the political economy of new media; blogs may provide an alternative form of "citizen journalism," but they do so in the context of a twenty-four-hour news cycle whose overall properties and topics are shaped in dialogue with CNN and Fox News. Marx commented that men make history, but not in circumstances of their own choosing; much the same can be said of digital media.

Culture and Technology

Geertz (1973, 448) once described culture as "stories we tell ourselves about ourselves." In their study of the Internet in Trinidad, Daniel Miller and Don Slater (2000, 11) invoke this image when they talk of the Internet providing people with resources that help them "deliver on pledges they have already made to themselves about themselves." Miller and Slater's study gives us some alternative ways to think about the relationship between technology and cultural practice, and therefore what a cultural account of information technology might be.

Their account is rich and multifaceted, but a central element focuses on how the Internet is implicated in the production of Trinidadian ("Trini") identity, particularly in the context of contemporary transnationalism. In the spirit of George Marcus's multi-sited ethnography (1995), Miller and Slater observe that the island can by no means geographically contain present-day Trini culture. Economic conditions force people overseas, especially to New York and London, where earning opportunities are better so that people can send money home to their dependents. The challenge that this poses to "pledges made to themselves about themselves" is that much of Trini identity is built around the notion of relaxed good times shared with family and friends. When significant numbers of families have nuclear family members living overseas, the kind of closeness and caring that play such a major role in Trini identity are difficult to achieve. The Internet therefore supplies a new means of enacting those aspects of Trini identity. A mother checks the weather forecast in London and reminds her daughter to take an umbrella to work; best friends spend time hanging out ("limin") online.

One could imagine how, in the hands of analysts less thoughtful, less grounded ethnographically, and less attentive to the nuance of cultural practice than Miller and Slater, this could be told as a story of the ways in which their Trini informants had appropriated a foreign technology and turned it to their own needs. This is not, however, the story that Miller and Slater unfold. Rather, they show that the Internet is encountered as always already Trini. There is no transformation or appropriation at work here because there is no initial alienation to be overcome or responded to. In Trinidad, the Internet is read as Trini as a matter of course; through those cultural lenses, the Internet comes naturally into focus as enactive of Trini needs and values.

Similarly, in the previous chapter we sketched aspects of how visions of ubiquitous computing are manifested in Singapore and Korea. We noted that the ideas of technology and society, ubiquitousness and spatiality, and effectiveness and efficiency, or nurturance and nature, presence and absence, and regulation and responsibility, intersect with technological agendas to produce different kinds of expectations and experiences than formed part of the animating (and U.S.) vision of technology.

Our goal is not to say, "Look how different things are over there" (even though, of course, they are), because the simple movement of technology from one context to another is not itself particularly of interest. Nor is our goal to say, "Look how the original (and naturalized) vision of ubicomp has been appropriated by these cultural forces," because of course the original vision of ubicomp was in no way naturalized and was itself deeply culturally inflected.

Instead, we want to highlight the ways in which, across many sites—Singapore, Korea, Trinidad, and the United States—technological imaginaries are deployed in support of cultural practice. The relationship between technology and culture being expressed here is one of information technologies as sites of social and cultural production—as points at and around which cultural values are enacted, produced, shared, reified, represented, and reaffirmed. If a taxonomic perspective—an appeal to anthropology's traditional orientation toward exotic others—has a role here, it is to defamiliarize us with the mundane aspects of our own cultural experience and so to help direct our attention to the shared practices of cultural meaning-making that underlie all acts of technological formulation and creation (Marcus and Fischer 1986; Traweek 1988).

The term "culture" obviously often obscures as much as it reveals, particularly as it has been taken up in engineering and design disciplines. Most commonly, it is used there in order to distinguish between a range of alternative sets of practices and expectations with an influence on information technology adoption and use. So questions of "cross-cultural design" or "adoption patterns in different cultures" draw on a taxonomic reading of culture in order to categorize people and groups according to a set of features of a given cultural setting.

Culture is anthropology's topic, and while anthropology also employed a broadly taxonomic view of culture until the mid-twentieth century, this perspective has for a range of reasons largely disappeared. Some of these have to do with the pragmatic problems of the taxonomic view—for example, where does one draw boundaries between cultural groups, what happens at those boundaries, how do we account for changes over time, and what does it mean to apply a notion of cultural separation in a globalized world? Other reasons are conceptual and political—that, for instance, cultural difference is generally defined relative to an ethnocentric norm, the categories adopted are usually not ethnographically warranted, people participate simultaneously in multiple cultural contexts, culture is performed and enacted rather than inhabited and received, and cultural categories are implicated in various forms of political and state oppression as well as regulation. The taxonomic view has given way to an alternative one that looks on culture as a generative, semiotic phenomenon. From this vantage point, culture provides a set of lenses through which everyday life is encountered and lived, and that give elements of everyday life meaning.

The Social and Cultural in Ubicomp

We have taken some time to set out these issues because they are central to our project here and the form of interdisciplinarity that this work represents. There are many configurations in which the social and computer sciences might be brought together in a collaborative enterprise.

Computer science might provide a set of tools and technologies to social science for conducting work, and perhaps gathering and analyzing data, or it might offer a domain to be subjected to some form of social analysis. Conversely, social science might supply some techniques to computer science for understanding how technology gets used, or why, where, or by whom, or it might provide a study of a setting into which computer technology could be introduced. Our goal here is a different one, though. We start from the position that technology and social life are not easily separated—that social life is already entwined with technologies and technological practices of all kinds, and that technologies and their designs are embedded in social systems that they reflect, reproduce, and transform. Accordingly, we do not break our topic down into technical versus social considerations; the two are neither independent nor indeed separate spheres of concern. We likewise do not adopt a concern with social shaping, in which digital technology is made consequent to social factors, or a concern with the social impacts of ubicomp, in which the march of technology is seen as inevitable and external to the social world. When we speak of the social and cultural aspects of ubicomp, we are attempting to set out a generative account of the practices that surround information technology, making it meaningful in everyday practice while operating as a site for the production of social and cultural realities and imaginaries.

What does this mean for the study of ubicomp? First, it means that we need to move beyond attempts to read culture taxonomically and focus instead on the essential processes of meaning making that constitute the technological experience. Rather than invoking culture as an explanatory mechanism ("Oh, they do that because it's part of their culture"), here we are interested in culture and its processes as a topic of inquiry.

Second, we outline how social science is already central, if sometimes implicit, in the ubicomp enterprise (or inversely, how ubicomp is only in part a technological enterprise). This entwining of concerns is deep and unavoidable. When someone writes of ubicomp purely from a technical perspective (or purely from a social one), it is not that the person has chosen to focus narrowly and exclude other concerns but rather that the other parts have been left entirely implicit—and, in our opinion, problematically so. An explanation of the social aspects of ubicomp that fails to account for the material conditions of hardware and software is not entirely solid; similarly, neither is any description of the technological aspects of ubicomp that fails to account for the social commitments, expectations, and imaginings at work in the technological enterprise.

The generative view that we have tried to outline here is precisely the source of this inextricable entwining. When technology is being used to "do" culture, when it is encountered as already cultural, and when it is encountered through a cultural lens—as it must inevitably be—then a form of interdisciplinarity is called for that does not simply manifest a division of labor that presumes and reproduces a separation between the concerns of each.

In this chapter, we have explored the tensions between these two competing notions of culture—one rooted in contemporary ethnographic and critical theory and the other based in the more instrumentalist practices of ubicomp. Drawing on examples from the anthropological canon and the authors' recent work, we suggest that the adoption of a more generative notion of culture within ubicomp has profound and destablizing consequences, which we discuss at length throughout the book. By taking topics such as mobility, privacy, and spatiality—topics central to any discussion of ubicomp's past, present, and future—and taking them as both technical and social considerations simultaneously, our goal is not only to illuminate the problems and practices of ubicomp systems but also to spotlight a new way of inquiring into them.

4 A Role for Ethnography: Methodology and Theory

It is in understanding what ethnography is, or more exactly what doing ethnography is, that a start can be made toward grasping what anthropological analysis amounts to as a form of knowledge. This, it must immediately be said, is not a matter of methods. From one point of view, that of the textbook, doing ethnography is establishing rapport, selecting informants, transcribing texts, taking genealogies, mapping fields, keeping a diary, and so on. But it is not these things, techniques and received procedures that define the enterprise. What defines it is the kind of intellectual effort it is: an elaborate venture in, to borrow a notion from Gilbert Ryle, "thick description."

—Clifford Geertz, The Interpretations of Cultures

As should be increasingly clear by now, ubicomp is unusual among technological research arenas. Most areas of computer science research—such as programming language implementation, distributed operating system design, or denotational semantics—are defined largely by technical problems and driven by building on and elaborating a body of past results. Ubicomp, by contrast, encompasses a wide range of disparate technological areas brought together by a common vision of computational resources deployed in real-time, real-world environments.

Realizing, or at least moving toward, this common vision has necessitated a blending of disciplinary approaches from computer science and engineering with some social science perspectives and practices. As we laid out in chapter 2, this mix could hardly be described as seamless. Indeed, given that ubicomp agendas have been pursued in industrial, academic, and governmental settings across a number of national arenas, a unified framework has never emerged. That is not to say that ubicomp lacks a point of view or clear directionality. The examination of recent publications, talks, grant proposals, and current research we referenced in the second chapter affirms that there are agreed-on as well as intellectually/

institutionally sanctioned areas of study and impact; if you are in ubicomp, it seems, you are doing systems, sensors, or users. Despite the emergence of a certain kind of canonical thinking within ubicomp, the doing of systems, sensors, or users has been characterized by an eclectic approach to methodology and theory.

In this chapter, we use the term "methodology" formally, to encompass not just the craft methods and techniques that a discipline employs to do its work—no matter how emblematic or charismatic they become—but also the epistemological foundations of the discipline, and the ways in which methods feature as part of a broader set of conversations about forms of knowledge production along with the kinds of objects that disciplines examine and create. Surveys, for example, are not simply a convenient way of sampling large populations; they are also reflections of a set of underlying commitments to topics of statistical relevance, objective measurement, generality rather than specificity, population comparisons, and the power of numbers to talk about people. Prototyping approaches similarly reflect a particular philosophical position on technologies and their portability, and the relationships between contexts, technologies, and practices. In this chapter, we wish to reconnect the ways we approach research questions (i.e., methodologies) with the ways in which such questions might be framed, articulated, and addressed (i.e., theory).[1]

This broader view of methodology and theory implies that there is more to the eclecticism of ubicomp research design practice than simply a "mix-and-match" approach that borrows techniques from different places. A concern with methodology and theory rather than method alone means that we need to understand the commitments involved in the various techniques that we employ, and the consequences of their combination. Brian Cantwell Smith (1996) uses the metaphor of commercial exchange to describe this caution to methodological syncretism: when you use an idea from somewhere else, he suggests, you need to be able to say what you paid for it, how you brought it home, and what kinds of damage it suffered along the way. Accordingly, it is appropriate that the happy eclecticism of a domain such as ubicomp be accompanied by some reflection on just what the nature and perhaps even histories of our methodologies might be, and what theoretical frameworks they imply or recall.

In this chapter, we explore this broader view of methodology and theory, using ethnography and its relationships to ubicomp as a starting

1. Here we want to distinguish between methods—the pragmatic tools of our trade—and methodologies—the framing conceptualizations of those selfsame tools.

point. This is hardly a new proposition. As we discussed in chapter 3, ethnography found its way into ubicomp early on. Weiser and his Xerox PARC interlocutors were influenced, accordingly to their own re-retellings (Weiser, Gold, and Brown 1999), by the likes of Suchman and her Work Practice and Technology Group. These early encounters help shape a ubicomp charter that, rhetorically at least, valued human relationships, social context, and what Weiser referred to as "the real world"—all theoretical and methodological concerns which with ethnographers, and indeed other social scientists, are familiar.

Within ubicomp, the adoption of ethnographic techniques has been associated with two trends. The first is the emergence of computer-supported cooperative work (CSCW) as an area of inquiry, which in turn placed an increasing emphasis on the social organization of activity, and hence on methodological approaches by which that social organization might be understood. (It was arguably through CSCW research that Weiser was first exposed to ethnographic methods and perspectives.) The second was the participatory design movement, arising especially in Scandinavia but with global influence. Politically, participatory design was strongly concerned with issues of workplace democracy and participatory involvement in the changes in working conditions implied by computerization; methodologically, it sought approaches in which member's perspectives were valued. For participatory design, ethnography may have been an expedient tool rather than an intellectually motivated approach, and indeed it has always stressed a pragmatic, multimethod approach. Nonetheless, through participatory design, CSCW, and allied perspectives, the use of ethnographic methods became more familiar to ubicomp researchers. They seemed to offer a means by which the complexity of real-world settings could be apprehended and a toolbox of techniques for studying technology "in the wild" (Grudin 1990).

Over the last twenty years or so, many researchers and practitioners in ubicomp have turned toward a broad array of social sciences and social science toolboxes,to find ways to understand this real world as well as the social contexts in which both users and technologies might be embedded. Social scientists have remained present in many centers of ubicomp inquiry: Intel, Xerox PARC, Georgia Tech, the Massachusetts Institute of Technology, the University of California, Lancaster University, the University of Nottingham, and Nokia, to name just a few. Ethnographic approaches are increasingly prominent as the means by which this might be accomplished. However, a wide range of forms of social investigation travel under the ethnography banner in ubicomp, suggesting that there is still considerable debate over what ethnography is and how it can best be employed in research, design, and deployment contexts. For the most part, ethnography has come to be regarded as a toolbox of methods, divorced from a larger set of theoretical and methodological concerns that give it form and rigor. Ethnography is too often seen as an approach to field investigation that simply generates requirements for systems development by providing a clear sense of "what users want." This is perhaps ironic given that most ethnographers cringe at the very notion of users—indeed, a great deal of ethnographic work in the last fifty-plus years has argued against this conception of the user (Sharp 1952; Pelto 1973; de Laet and Mol 2000).

As mentioned in chapter 3, ethnography has had its own histories and canons, and it is located within the broader trajectories and debates of the social sciences. To understand the ways in which ethnography figures within and without ubicomp is to understand not just its methods but also its methodologies and larger epistemological concerns with questions of reflexivity, voice, stance, and standpoint—most of which are largely absent from ubicomp practice. It is valuable to step back and consider what happens when two disciplinary, conceptual, and methodological approaches come together, and how the relationship between them could be articulated. Thus, in this chapter we explore the relationships between ethnography and ubicomp, beginning with a critique of how ethnographic accounts function within ubicomp. Here, again echoing our earlier work, we look to the "implications for design" practices as the signal manifestation of ethnographic knowledge within ubicomp.

Throughout this chapter, we argue that by relegating ethnographic knowledge to implications for technological design, ubicomp practitioners fail to capture the value of ethnographic investigations, insights, and knowledge. Yet it does function as a useful avenue to open up a larger conversation about not only how ethnography is currently prefigured in ubicomp but how it could be powerfully reimagined, too. We are particularly interested in how ethnographic theory can help reposition research questions and directions without a reliance on fieldwork. This chapter thus illustrates the implications for design that might be derived from classical ethnographic material, and it shows that these may not be in the form that ubicomp research normally imagines or expects.

Ethnography as Implications for Technological Design

As intellectual disciplines develop, genre conventions emerge, shaping their research designs and outputs. In interdisciplinary areas such as ubicomp, early work in the field tends to be highly divergent in method and approach, as practitioners—individually and collectively as a field—attempt to find ways to combine perspectives, conceptual frameworks, and methods. Finding an appropriate balance between theory and practice, determining broadly agreed-on metrics for success, and developing common vocabularies for the problems and phenomena of study are all means by which, over time, common consensus about research is developed. Scientific disciplines are normative enterprises, where the process of peer review tends to encourage conformity to a core set of values and approaches (Campbell 1969).

This process can be seen in the research papers produced in a field. Charles Bazerman (1988) has detailed the ways in which transformations in the structure and tone of scientific publishing accompanied the transformation of the conduct of science itself, reflecting its increasing professionalization; the process of ensuring conformity to documentary standards is part of the "boundary work" by which disciplinary limits are maintained and the boundary between "science" and "nonscience" is sustained (Geiryn 1983). Case studies illustrate how these conventions shape the development of scientific arguments and publications (Frost and Stablein 1992). Unsurprisingly, then, as ubicomp has matured and developed a sense of its own disciplinary identity, conventions have arisen regarding how we conduct and describe our research.

Here we want to focus on one of these genre considerations: the notion of implications for technological or information systems design, known in shorthand as "implications for design." We are interested in this notion as a matter of both research presentation and research construction. There is a great deal of tacit pressure on ethnographers and other social scientists working within the ubicomp context to generate implications for design. Indeed, any canonical paper reporting ethnographic field results in the ubicomp context will close with a section titled "Implications for Design." This section may be long or short, comprising discursive prose or brief, bulleted items, but it nonetheless figures as a stable feature of ethnographic reports. Informal evidence seems to suggest that the absence of this section tends to be correlated with negative reviews and rankings of the paper. A common lament to be found in reviews of ethnographic work is "Yes, it's all very interesting, but I don't understand its implications for design" or, somewhat more subtly (and intriguingly), "This paper does not seem to be addressed toward the ubicomp audience." The pressure to generate implications for design happens in both academic and industry contexts.

Two things are clear to us. First, the focus on implications for technological design is misplaced, misconstruing the nature of the ethnographic enterprise. Second, given this, it misses where ethnographic inquiry can provide major insight and benefit for ubicomp research. We are interested here in the politics and consequences of the manner in which the implications for design arise as a primary mechanism of ethnographic research. In framing it as a problem, we want to explore how implications for design may underestimate, misstate, or misconstrue the goals and mechanisms of ethnographic investigation.

Charting a New Relationship between Ethnography and Ubicomp

In what follows, we examine these questions by dealing in turn with four issues that arise around the problem of implications for design: the marginalization of theory, power relations between disciplines, a restricted model of the relationship between technology and practice, and the problems of representation and interaction. Some of these concerns could be classified broadly as "the politics of representation," while others could be categorized as "the politics of design." Certainly, the considerations are political in a number of ways, and we return to some overtly political issues at the end.

In particular, we argue against the idea that ethnography is undertaken in order to uncover such implications, in the narrow sense that requirements capture. That position is based on a view of ethnographic work as purely empirical, as a process of going out and finding facts lying around in the world, dusting them off, and bringing them home to inform, educate, and delight. We suggest that there are four considerations that get lost if we concentrate purely on ethnographic research-generating implications for design. First, we must recognize the theoretical work of ethnography, or the fact that ethnography is an interpretative, analytic practice. Ethnography's commitment to the production of social facts in culturally organized settings in fact requires this, and it also necessitates that the work of the ethnographer is more than simply collection. Second, there are disciplinary power relations at stake, by which ethnography is here placed in a service relationship—just the sort of relationship that designers have been careful to avoid in their own work, and for good reasons—and this relationship also implies a specific and problematic location for agency within design. Third, to the extent that ethnographic work centers on the ways in which people produce as well as enact social and cultural settings, the implications-for-design model inappropriately emphasizes technology over practice as we set about understanding the interplay between the social and technical. Fourth, it is important to pay attention to how ethnography in ubicomp can be used to limit, rather than expand, the engagement of users in design practice, arguably recapitulating some of ethnography's history in colonial state enterprises (as we discussed in chapter 3) and so prompting a good deal of resistance from practitioners grounded in anthropology's disciplinary history or concerned with the politics of representation.

We draw on three insightful explorations of the problems of ethnography and design in different contexts: Robert Anderson's examination of the issue of ethnography and requirements (1994), Mark Ackerman's reflections on the social-technical gap (2000), and Graham Button's comparison between different models for ethnographic analysis (2002). They help to illuminate a complex and intricate set of disciplinary relationships, which will be addressed here through the four interrelated topics, starting with the question of the marginalization of theory.

The Marginalization of Theory

As outlined above, ethnographic methods were originally brought into ubicomp research in response to the perceived problems of moving from laboratory studies to broader understandings of the social organization of settings of technology use. It might be more accurate to say that in ubicomp research, ethnographers were adopted rather than ethnography itself. That is, a number of social scientists making use of ethnographic approaches turned their attention to questions about the interactive technologies community and found a positive reception for aspects of their work (Nardi 1993; Suchman 2007; Sproull and Kiesler 1991). This distinction is important because as ethnographic approaches have gained more visibility and currency within ubicomp, some problems have attended the ways in which ethnography has been understood.

In particular, the dominant view of ethnography is that it provides a corpus of field techniques to ubicomp researchers for collecting and organizing data. The term "ethnography" indeed is often used as shorthand for investigations that are to some extent in situ, qualitative, or open-ended. We have both read and reviewed papers where "ethnography" was used to mean that the researcher had spoken to a test subject outside the context of a usability lab. Similarly, the term is frequently used to encompass specific formulations of qualitative research methods such as contextual inquiry (Beyer and Holtzblatt 1997). So here, the defining characteristic of ethnographic investigation is taken to be its spatiotemporal organization—the ethnographer goes somewhere, observes, returns, and reports—or what Button refers to as "scenic fieldwork" (2000, 330).

Perhaps unsurprisingly, this reading of ethnography has often been aligned with the requirements-gathering phase of a traditional software development model. Laboratory methods can provide certain kinds of answers to certain kinds of questions that can shape the design of a software system. By analogy, ethnography is usually conceptualized as a set of field techniques that can supply different sorts of answers to different sorts of questions—especially questions about technology in everyday settings—that nevertheless will stand in much the same kinds of relation to design exercises. The same empirical urge can be seen in the adoption of other approaches such as cultural probes and related approaches, even though probes were not designed as data collection instruments (Gaver, Dunne, and Pacenti 1999; Hutchinson et al. 2003; Boehner et al. 2007).

This view of ethnography as purely methodological and instrumental supports the idea that implications for technological or information systems design are the primary or even sole output of ethnographic investigation. From this perspective, the reason to adopt ethnographic methods is not that it will generate quite different kinds of understandings from laboratory investigations but rather that laboratory approaches are methodologically unsuited to the target domain.

In reducing ethnography to a toolbox of methods for extracting data from settings, however, the methodological view marginalizes or obscures the theoretical and analytic components of ethnographic study. Ethnography is concerned with the member's perspective and experience, but it does not simply report what members say they experience. Even in ethnomethodological ethnography, which rejects sociological theorizing in favor of explicating observable practice, ethnography makes conceptual claims; it theorizes its subjects, even if the theories presented are the subjects' own (Button 2000). To the degree that ethnography presents not simply observations but also relationships between observations, it is inherently interpretive. Indeed, ethnography's outputs are frequently not analytic statements purely about members' experiences but instead about how members' experiences can be understood in terms of the interplay between members and the ethnographer.

Anderson (1994) insightfully explores the relationship between ethnography and requirements, paying particular heed to the way in which what we have called the "methodological approach" consistently marginalizes or obscures the analytic component of ethnography—and, importantly, how in doing so it both underestimates ethnography and fails to realize its potential. Anderson draws attention to three considerations.

First, he notes that ethnography must be seen primarily as a form of reportage. It is after all ethno*graphy*—a form of writing and a way in which a cultural understanding is inscribed as a literary form. Writing, then, is central and the ethnography is not itself the project; the written form is its final outcome. Consequently, we must pay considerable attention to its rhetorical form and construction. Much contemporary debate around ethnography has been animated by a close look at ethnographies as texts, to how they implicitly or explicitly construct the roles of author and reader as well as the object of inquiry (Clifford 1983; Clifford and Marcus 1986; Geertz 1988). Ethnographies are not mere acts of writing up "user reactions" or focus group discussions, or transcribing interviews; they are instead representations of the world that the ethnographer encounters.

Second, Anderson observes the role of particular rhetorical strategies, not least the juxtaposition of strategically chosen exemplars, such as, in one of Anderson's examples, patterns of sharing customizable software as explored by Wendy Mackay (1990) and the marriage practices of the Bororo as detailed by Claude Lévi-Strauss (1969). Despite a certain ethnographic tendency to operate as "merchants of astonishment" (Geertz 2000), the goal of such juxtapositions is not merely to dazzle and surprise; rather, it is to reveal certain underlying logics of social practice. Once more, this is fundamentally an analytic move. What is revealed is the conceptual organization of cultural settings, and while the goal is to reveal and explicate as opposed to create, the ethnographer is far from a passive agent in the production of this organization as a research outcome. However, creating moments of surprise or astonishment in which the audience/readers are forced to challenge their current framing of a situation does not always sit comfortably within a research structure that favors stable questions and research problems. Questions are answered and problems are solved or addressed. The notion that it might be the wrong question or an inappropriate problem statement seems to fit poorly within the ubicomp tradition.

Third, Anderson emphasizes the reflexive character of ethnographic analysis. This means that ethnography is not only about the culture under study but equally, implicitly or explicitly, about the cultural perspective from which it is written and that of the audience to whom it is presented. By telling an ethnographic story about some Other, the ethnographer also tells a story about ourselves (Marcus and Fischer 1986). How many ubicomp papers or presentations account for the author's stance? Such moves, when they do happen, are greeted with skepticism and even moments of hostility. Yet the question of subject position is important here. Ethnographic data are not unproblematically extracted from a setting but rather generated through an encounter between that setting and the ethnographer. Students learning ethnographic methods for the first time, especially those from positivistic scientific traditions, frequently express the concern that the ethnographer, as an instrument, must inevitably distort the data and introduce an element of uncontrolled subjectivity in contrast to alternative approaches. Paradoxically, the situation is in fact reversed. Quantitative and survey techniques depend on subjective judgments about the categories of observations that remain implicit in the data (Becker 1993; Garfinkel 1967), but by contrast ethnographic methods explicitly require that the ethnographer incorporate the context of the social relationship between ethnographer and subject or setting. So, for example, ethnographic understanding depends critically on recognizing that the view of the setting (or the interview responses) that one gains is inevitably shaped by ones' subject position—ethnic, sexual, or class markers; access to resources and power; introduction and social position; and so on. One way in which the methodological view of ethnography practiced in ubicomp often marginalizes or obscures the analytic component of ethnographic investigations is to cast the ethnographer as a channel for the relatively straightforward movement of data from the field to the design studio. As Diana Forsythe (1989) tellingly comments, an ethnographer is not a tape recorder.

Power Relations

The second consideration illuminated by the problem of implications for design is a more broadly political one, concerning the relationship between the constituent disciplines in ubicomp.[2] The particular issue we explore is how the idea that the goal of ethnography is to generate implications for design construes the disciplinary relationship. There are three concerns here. First, the implications-for-design model postulates design as the natural end point of research inquiry and, therefore, designers as the

2. It is hard to deny the power differential between engineering sciences and social sciences in terms of academic and funding structures; a brief perusal of the relative size of research grants will demonstrate that amply. This disparity has consequences both large and small. At a large scale, it creates a status hierarchy in which engineering demands tend to override social ones; at a small scale, it results in an imbalance in participation in scientific meetings (since social scientists are rarely in a position, for instance, to fund their own travel to program committee meetings and conferences, as venues like the annual conferences for ubicomp and HCI normally demand). Despite these huge practical obstacles, we focus here on some more conceptual concerns.

gatekeepers for that research. Second, in doing so it places ethnography outside the design process itself. Third, it places those whom ethnographers study outside the design process too. The third consideration is one that we will examine later, but the first two are more immediate topics.

The question at stake here underlies *any* interdisciplinary effort: the difficulty of achieving a true synthesis or mutually constituted discursive arena, rather than degenerating to a case in which one discipline is essentially in service to the other. Certainly this is commonly understood in computer science; as computation has become an increasingly critical element of other scientific enterprises, computer scientists are wary of becoming programmers in service to other disciplines.[3]

Clearly, in this case the issue is that technological or information systems design is the tail that wags the dog. The distinction to be drawn is perhaps that between user interface design and HCI as domains of study. If the interaction between people and computers—or between people through computers—is itself a domain of inquiry, then the call for ethnographic studies to deliver implications for design is somewhat disingenuous, especially perhaps at conferences with titles like Human Factors in Computing Systems (as the CHI conference is more formally named), or Ubiquitous Computing rather than Human Experiences of Computing Systems or Designing for Ubiquitous Computing. It instead suggests that ethnographic investigations (indeed, ubicomp research studies) are relevant only inasmuch as they support technological design (and not simply in terms of helping to understand HCI). While it is obviously important, in a design- and technology-oriented field, to be concerned with highlighting and correcting problems in current technologies, for a range of reasons ethnography is not necessarily best oriented toward the creation of new sorts of technological or consumer artifacts. Sometimes, after all, the most effective outcome of a study might be to recommend what should *not* be built. More to the point, an analysis of the cultural and social organization of some specific setting or occasion is often best articulated independently of *specific* systems, technologies, or design opportunities.

Returning for a moment to Miller and Slater's study of the Internet in Trinidad (2000), the power of their analysis does not lie in specific recommendations about the ways in which technology might be best designed

3. At a recent meeting of the recipients of a particular program of interdisciplinary research grants, this was a major source of tension and frustration. The fascinating solution was to advocate what was called "vertical interdisciplinarity"—interdisciplinary engagements between computer scientists of different stripes.

to fit into a Trinidadian context; it lies in their critique of the ways in which the domains of "natural" and "virtual" worlds are conceived as well as argued through information technology. Miller and Slater demonstrate how the technology does not create a place outside everyday life but rather provides a new platform on which everyday cultural experiences can be performed. They show how the Internet supplies Trinidadians with another way of "being Trini"—indeed, ways of being Trini that the practical realities of daily life may imperil. What Miller and Slater question is the conventional separation between virtual and real domains; the Trini experience of the Internet, though, is one that is coextensive with, and indeed grows out of, Trini experiences of everyday life.

This calls into question a number of the assumptions that lie behind the notion of implications for design from ethnographic work. First, who is doing the technological or information systems design in these scenarios? There are at least three potential design actors here: the ethnographers, the technologists, and the people themselves. A particular set of relationships between these constituencies is postulated by the traditional focus on implications for design (especially that a designated and demarcated group of designers are empowered to perform design, of which others are passive consumers). Second and perhaps more problematically, it causes us to reconsider just what design looks like—the technology itself, or the form of its local adaptations and appropriations in particular social and cultural contexts. Third, by focusing on specific designs as the point at which ethnographic and technological considerations meet, are we doing justice to the ethnographic perspective, and are we getting the best technological outcomes? At what point can ethnographic contributions have their greatest impact on technology development and deployment? Kjeld Schmidt (2000) claims that the most influential workplace studies in CSCW have been ones that did not harness themselves to specific design efforts or limit their discussion of implications to then-available design opportunities. Fourth, and consequently, is the success or value of an ethnographic investigation best determined by what design decisions it can support or by what forms of learning it might enable. Or to put it another way, what forms of knowledge can ethnographic studies generate?

Technology and Practice

Following from some of these questions, we examine the relationship between technology and practice postulated by the implications-for-design approach. In particular, as discussed above, we highlight two assumptions implicit in this approach. First, it constructs ethnography as a point of mediation between, on the one hand, a domain of everyday practice and, on the other, a domain of technological design. Second, it implies that people will encounter technology as something just as it was designed and, hence, is appropriated or incorporated into practice. Each of these assumptions is problematic from the ethnographic perspective.

Ackerman (2000) provides the metaphor of the "social-technical gap"—essentially the gap between our technological reach in the design process and the realities of technologies in practice. In drawing attention to this, he spotlights the notion of design as a bridge. This shows how through a range of methodological innovations (such as, perhaps, the incorporation of ethnographic methods alongside controlled laboratory studies), ubicomp has sought to narrow the gap or to bridge it. Ackerman critiques the intuition that people appropriate and adapt technologies because the technologies are poorly designed and that better-designed technologies would obviate the need for such adaptation and appropriation.

By contrast, ethnographic approaches yield a different perspective on the creative processes by which people put technology into practice and meaning. These are seen as consequences of everyday action, not as a problem to be eliminated. Technology here is a site for social and cultural production; it provides occasions for enacting cultural and social meaning and, as with technology, so also with space, gender, family, time, animals, food, death, emotion, and everything else. Seeking to close the gap through the application of ethnographic methods is a contradiction in terms; the gap is where all the interesting stuff happens, as a natural consequence of human experience. The gap between what people say they are doing or will do in the future and the actions they are currently undertaking is fertile ground for ethnographic inquiry. Design is critical, but designs must always be put to work in particular contexts, adopted and adapted by people in the course of practice.

In this way, the domain of technology and that of everyday experience cannot be separated from each other; they are mutually constitutive. The role of ethnography, then, cannot be to mediate between these two domains because ethnography does not accept their conceptual separation in the first place. By introducing and focusing on the notion of the gap, Ackerman suggests not that it is the fundamental problem to be solved but rather that it is the fundamental phenomenon to be understood.

It is lived and embodied practice—the articulation of aspirations and cultural ideals along with all the spaces in between—that gives form and meaning to technology. The focus of ethnography is the ways in which these bring technology into being. From this perspective, and drawing again on the notions of reflexivity raised earlier, we might suggest that what ethnography problematizes is not the setting of everyday practice but instead the practice of design.

Certainly, though, it refigures users not as passive recipients of predefined technologies but rather as actors who collectively create the circumstances, contexts, and consequences of technology use. Ubicomp research has long had an interest in aspects of how people might configure, adapt, and customize technologies (e.g., Dourish and Button 1998; MacLean et al. 1990; Nardi 1993; Bell 2006a). Still, this ethnographic view does not simply focus on how people explicitly transform or program interactive technologies; it looks at how those technologies take on specific social meanings through their embedding within systems of practice—systems of practice that might encompass more forms of technological engagement than a traditional focus on use provides (Satchell and Dourish 2009). As a focus of attention on ubicomp research, design in this sense goes beyond giving form to technologies to encompass appropriation—the active process of incorporation along with the coevolution of technologies, practices, and settings.

Broadening the Scope of Ethnographic Impact in Ubicomp

As the previous sections suggest, ethnography's analytic contributions do indeed have profound implications for design, but these implications go beyond the laundry list of features and considerations that are often requested. Our resistance to a bulleted list of requirements comes partly from the fact that they underplay the more radical implications that may be caught up in ethnographic work; if the ethnographer returns from the field with little more than the lesson that the object in question should be green, fit in a handbag, and run for at least three weeks on two AA batteries, then we might venture that there is not much to the ethnography.

Far more ethnographic work is potentially relevant for design, whether or not it was conducted in a design context or in relation to new information, communication, and entertainment technologies, or even if it lacks an implications-for-design section somewhere in its closing pages. Yet we would argue that such ethnographic inquiry can be extremely influential for design without requiring the conventional implications-for-design section. In fact, implications for design that emerge at the time of the ethnographic inquiry have inherently short shelf lives and so may obscure more lasting contributions (Dourish 2007). Perhaps the most useful strategy when engaging with ethnographic work is to "read for theory" as much as for empirical evidence, since in the end these may be where the truly significant implications lie.

In what remains of this chapter, we illustrate two recent areas of design-oriented research and the ethnographic work that could inspire and shape them. The two are already areas of current research and design attention within ubicomp: affective computing and mobile computing. The ethnographic work on which we want to draw, though, was conducted well outside the technology domain, and much of it more than two decades ago. What we want to illustrate are the profound implications they hold for technological and information systems design.

Affect

The traditional focus of both ubicomp and HCI on the cognitive aspects of interaction design has recently been supplemented by a range of new perspectives that look beyond the purely instrumental aspects of interaction. One of these perspectives has centered on emotion and affect, as developed most particularly by Rosalind Picard (1997) and Don Norman (2004). Both of these authors argue that the traditional focus on task performance has been overly reductive, modeling people in purely computational terms and neglecting other important aspects of experience. Cognition is not disembodied and disconnected from other elements of human experience; a significant body of work highlights the role that emotion plays in decision making and other areas of cognitive activity. Accordingly, research in affective computing has begun to investigate the possible relationship between ubicomp and HCI analytic-and-design practice and the affective aspects of interaction. Among other topics, affective computing researchers are investigating whether we are able to build systems that model and respond to a user's emotional state and then to be able to craft responses and design interactions that take that state into account—for instance, by attempting to recognize and defuse stress. This work places the emotional aspect of interaction alongside the more traditional cognitive and analytic elements.

As we have contended elsewhere (Boehner et al. 2005), there is a curious irony at work in this research. On the one hand, affect is rhetorically figured as an alternative or supplement to cognition; the claim is that we have placed all our attention on one element of human experience at the expense of others and so we need to redress the balance. On the other hand, at the same time as this opposition is presented, affect is figured as a concern of the same order or type as cognition. Like cognition, affect frequently appears in this research as a private experience, as something individual that is internal and closed off from the world. Affect is something that lies, both temporally and spatially, between perception and action. While turns to affect as an important interaction modality attempt to throw off the shackles of pure cognitivism, they seem to carry a significant amount of that legacy with them anyway.

Ethnographic studies of emotion can provide an alternative account that is useful in two ways. First, it shifts us toward a different way of imagining the relationship between information technology and affect, providing a different set of design strategies. Second, it highlights the cultural specificities of this parallelism between emotion and cognition.

Here, we use two ethnographic accounts of emotion: Catherine Lutz's study (1986, 1988) of everyday emotion in Micronesia, and Lila Abu-Lughod's study (1986) of emotional expression among the Bedouin. Other studies—such as Fred Myers's work (1979, 1986) among the Pintupi in Australia or Michelle Rosaldo's work (1983) among the Ilongot—are also relevant, but we will limit our discussion here to these two.

Abu-Lughod (1986) offers a detailed ethnographic account of honor and modesty among the Alwad 'Ali, a group of Bedouin tribes of Egypt's Western Desert. Much of her exploration turns around questions of gender and kinship, and the code of honor as it is entwined with these. What is most relevant here is the issue of emotional performance. Modesty and emotional reserve are hallmarks of conversation and interaction among the Bedouin, for whom a code of modesty results in an outward stoicism. What intrigues Abu-Lughod, however, is the distinction between the reserve of everyday speech and the emotion expressed in short fragments of poetry that the Bedouin might mutter, sing to themselves, or casually drop into conversation in the course of daily life. These brief, haiku-like fragments of poetry are often laden with joy, sadness, and longing, even as the people who utter them remain stoic and passive. What is more, these fragments are seen by the Bedouin as a truer window into the soul than everyday deportment. Emotional performance is thus a way in which the code of modesty is maintained and enacted—a code that is itself strongly oriented toward gender performance and relations, being connected to the code of honor by which masculinity is defined and tested.

In Abu-Lughod's account, emotion is not treated as a thing apart from other aspects of social life, a purely private experience on which sociality is layered, but rather as a fundamental element of social and cultural reality. It is a way in which this social and cultural reality is performed and enacted, brought into being and maintained through specific emotional performances. The emotion is not a precursor to action; emotion, as a cultural object, is produced through concerted action.

Lutz (1986, 1988) is also concerned with emotion as a cultural category, which she unpacks using material from her time on Ifaluk. Her concern in the field materials is not simply with a different set of emotions that are expressed in other places but instead with a different way of thinking about emotion as a cultural category, a different role for emotion, and a different evaluation of its meaning. One of Lutz's extended examples is the concept of *song* (justifiable anger) among the people of Ifaluk. While Western societies frequently class anger as antisocial, *song* plays a distinctively prosocial role on Ifaluk; the danger of invoking *song* in others might be used to caution children to play quietly and not disturb their elders, or it might otherwise curb those behaviors that upset the balance of everyday life. The conditions for bringing out *song* in others are those that reflect cultural practice—so noisy play might certainly provoke *song*, but so might actions that display an inappropriate disregard for familial or ritual responsibilities, such as how food should be shared. To be able to identify and respond to one's experience as anger (rather than, say, other emotions or pains, such as an upset stomach), then one must be able to understand the world as offering the conditions for anger—conditions that are themselves cultural. *Song*, as a personal experience, is nonetheless a cultural product; cultural meaning generates the landscape of emotional experience.

In reading Lutz's account, it is critical to recall again two potential readings of the word "cultural" that we flagged in chapter 3: the taxonomic and the generative. The taxonomic reading is one that seeks to classify and categorize people as well as their attitudes according to cultural (frequently ethnic or national) traits, habits, or inclinations. As we argue in the previous chapter, it is this taxonomic reading of culture that is invoked when people attend to the different positive or negative associations that people from different parts of the world might have toward colors, or when they account for differences in technology use according to whether people come from individualistic or communitarian backgrounds. If we were to read Lutz's concerns in light of the taxonomic view of culture, then we would take it to say that people experience and express different emotions depending on their cultural background, or that the categorization and evaluation of different emotions is one that varies with culture. This may be true, but it does not take us very far and, what is more, it rests on a definition of culture that raises more questions than it answers (Dourish 2006b; Ortner 1984; Yengoyan 1986).

Lutz's account instead draws on a generative account of culture—where we see cultural understandings as lenses through which everyday life is experienced and interpreted. By this view, emotional experience is a consequence of cultural embeddings. Biophysiological events are interpreted according to cultural scripts. Whether I can make sense of my experience as that of anger relies on an existing set of cultural understandings—an interpretation of the events around me as those to which an angry response might be justified. Culture is what helps me tell the difference between anger and indigestion; it is generative of the experience. Critically, then, such putatively private aspects of experience such as emotion are always already cultural; cultural aspects of interaction are prior, not consequent, to perception and action.

In this we see the link back to Abu-Lughod's attention to the performative aspects of emotion and the ways that emotion is a site at which cultural realities are enacted. Further, Lutz argues that emotion is a key master category in Western thought—one that lines up with and is linked to other critical distinctions around which our thinking is organized, particularly in its distinction to cognition and rationality. So rationality is of the head, but emotion is of the body; rationality is controlled, but emotion is uncontrolled; rationality is cold, but emotion is hot; rationality is male, but emotion is female.

Neither of these studies were written in a technological context or for a technological audience, and neither provides a series of implications for design.[4] That is not to say that they do not have implications for technological design or that they do not indeed raise profound questions for how emotion is figured as a facet of technological interventions. These accounts demonstrate a nonessentialist characterization of emotion, in which the shaping of an emotional "landscape" is culturally determined. They also depict emotionality as an outcome of engaged cultural practice rather than as a precursor to action. Emotion is produced and enacted in socially and culturally organized occasions. Note that this is not simply an argument that emotion is playacting or pretense. To observe that a setting is culturally organized is not to suggest that it is false; nor are moments of solitude any less culturally organized than those of intense social interaction. Throughout these studies, enactment—the continual

4. Given that the studies were both published in 1986, any attempt to distill technological implications, if those had even been a topic, might seem absurdly dated in the context of contemporary computational developments. The temporalities of social and technical phenomena are important considerations in general when thinking about disciplinary relationships.

and ongoing production and reproduction of aspects of social reality—is a fundamental consideration.

Furthermore, these accounts suggest that emotional expression is a point at which cultural values are expressed and performed. Rather than thinking of emotionality as being shaped by cultural variables, they hold that emotional performance is itself a site of cultural production. They help to account for the ironic relationship between cognition and emotion in ubicomp and HCI discourse by demonstrating how these are aligned within broader category systems. This allows us to think past the representationalist point of view.

Emotion is therefore interactional as opposed to representational. This conclusion does not simply *raise* implications for design; it *is* an implication for design. Kirsten Boehner and her colleagues (2005) show the ways that this implication was worked through in the design of a system called Affector, which tackles the opportunities around affective computing from a nonrepresentationalist stance—one that supports the enactment of emotional sociality rather than attempting to uncover the parameters of an emotional model that underlies and shapes human action. Affector is a video communication system in which image-processing techniques are used expressively to transform real-time video images in line with affective states but without any representational substructure. Much as in Karl Weick's (1995) exploration of the aphorism "How can I know what I think until I see what I say?," Affector encourages participants to play until they "see how they feel." The system's display is not a presentation of an emotional state; it is directly a performance of one. If we think of emotion in terms of performativity, enactment, and cultural production, we are lead to a radically different way to conceive of affect in interaction.

Mobility

A second domain of recent interest in interactive system design is mobile technologies. Accompanying the increasing prominence of mobile telephones as interactive and computational platforms as well as the spread of wireless networks enabling mobile access to information, ubicomp has increasingly addressed problems of information access "on the move." Some of the problems here are simply the constraints that might be imposed in these settings—limitations on input devices, output devices, and computational power. Some are those that speak directly to the relationship between devices and the contexts in which they are deployed, focusing on the contextually appropriate delivery of services or information. Others are concerned with the settings through which mobile devices might move and how these spaces and spatialized resources might be made navigable and accessible to the users of mobile devices.

While the first category of applications is concerned largely with devices and their affordances, the latter two are concerned instead with the nature of location, movement, and spatiality—how it is that people orient, individually and collectively, toward the spaces we inhabit. One of the central concerns, then, is what locations mean from a human-centered perspective. The traditional approach is a cartographic, Cartesian one in which space is understood as a manifold that can be indexed by a coordinate system, even though that coordinate system might be hidden behind a more human-oriented system of labels (allowing people to navigate via terms like "home," "office," and "store" rather than opaque latitude and longitude).

Again, we might turn to ethnographic investigations to gain a different view of space—one that focuses on an understanding of space as it arises from within particular cultural practices, looking at topics such as migration, nomadism, tourism, or globalization. Each of these topics is clearly founded on some perspective on space and its meaning and, again, they provide us with a different lens through which to examine questions of people and movement.

One piece that exemplifies this approach is Nancy Munn's ethnographic work (1996) among the Warlpiri peoples of the central and western Australian desert. The relationship between people and the land in this particular Australian Aboriginal belief system is a complex one. The form of the contemporary landscape is the result of mythical creatures' actions in the "Dreamtime," a mythical period after the creation of the world but before the arrival of people.[5] Since these creatures stand in totemic relationships to tribes and clans of the contemporary peoples, the activities that can be "read off" the landscape also result in a series of ritual responsibilities and relationships to parts of the land according to patterns of kinship and lineage. The relationship is more than simply one of environmental stewardship; the landscape is the source of Warlpiri identity and law.

Furthermore, this binding of people to landscape is a continual one, maintained and renewed through the ongoing relationship of Dreaming (see also chapter 5). One's responsibility is to Dream the land into

5. The notion of the "Dreamtime" that figures in many accounts of Australian aboriginal culture and social systems is a contested one. Patrick Wolfe (1991) offers a layered analysis of the origins of that term, the ways in which it is variously deployed and resisted, and its hegemonic effect—suggesting a universality of experience rarely found in other accounts of Aboriginal Australia.

existence; it is through this Dreaming that the connection between people and the world is maintained and honored. This ongoing link is reinforced by the land also being seen to carry the resonances of human activities and events there as well as mythical events. So patterns of habitation and settlement, migrations, meetings, battles, and births and deaths also leave their impact on the land. The Warlpiri experience of the landscape is thus a cultural one. The topography of the land is, at the same time, encountered as physical, mythical, and historical.

Munn is especially concerned with spatial interdictions—the circumstances and conditions under which people are ritually excluded from spaces. For instance, the separation between women's and men's ritual practices (or "business") is based not simply on events but also on spaces; one will avoid being in the places where one might see or accidentally encounter the ritual events from which one is tabooed. Similarly, spatial taboos may exist between classificatory groups. One example is that between mothers-in-law and sons-in-law—not only those who are actually related by marriage but rather those who are the members of subsections for whom kinship rules dictate the potential to stand in this relationship. So as a practical matter, a spatial interdiction exists relating to the parts of town where the people from the relevant subsection cluster (Bell 1983).

These interdictions are manifest in various ways, most particularly in the detours that characterize much Aboriginal navigation, as people move through the landscape in ways that respond to the various characters of the topography. Munn (1996, 449) is concerned with "spatial prohibitions as a mode of boundary making"—that is, with how the forms of prohibition to which one is subject, and one's orientation toward them, are means by which the environment's organization is not just marked but produced as well. Sacred sites, with their historical and mythical resonances, are a source of these prohibitions, as specific ritual sites, events, and seasons. A further complication lies in the fact that spatial prohibitions may be tied to events and actors that are themselves mobile. Rituals move; people move; as they do, the locales from which one might be excluded move too.

Munn underscores the fact that these spatial exclusions are not marked by boundaries in the ways that we might imagine Western land claims to be defended or regions protected. The spatial model here is one of centers of ritual potency that resonate out into the environment. As she notes, the radius of power is not clearly delineable. Moving too close may bring bad luck, illness, or death, but "too close" is relative; it may be linked to seniority or kinship. One's knowledge of and relation to these centers of potency is culturally embedded. Added to this is a further complication—knowledge of country, sacred sites, and spatial exclusions is not universally shared by all Warlpiri. It is revealed only if and when it is necessary to do so. Even Warlpiri "maps" of Warlpiri country, when they are drawn, are always partial, contextual, and transitory. This is not cartography as we would recognize it in the Western tradition. The model of place at work here is cultural and relational.

Indeed, one of the reasons that the Warlpiri case is such an interesting one is the fact that in Australia, two completely different systems of spatial knowledge production rub up against each other in troublesome ways. Debates over the protection of sacred sites and Aboriginal land rights have been so bitter and so prolonged, not least because of the fundamentally different systems of spatial knowledge and reasoning at work (Turnbull 2000; Verran 1998), and the legal and legislative outcomes concerning native titles reflect some of the inherent contradictions of reconciling the incommensurable (Hill 1995).

A second ethnographic example—Liisa Malkki's work (1992, 1995) on national identity among refugees in Tanzania—reveals a different set of cultural concerns over movement and mobility. Malkki's fundamental concern here is "rootedness." In the context of globalization, large-scale transnational migration, and interconnected labor markets, anthropologists have long recognized that the objects and topics of their inquiry are not fixed in specific places but rather move around and take their shape within the world system, and that ethnography therefore needs to be multi-sited itself and to engage with multi-sited phenomena (Marcus 1995). These issues, however, are more directly present in Malkki's work, which looks at the ways in which national identity and rootedness manifest themselves for transnational migrants and refugees. She argues that the very figuring of rootedness and the authenticity of the indigenous reflects what she calls a "sedentarist metaphysics"—a notion that staying put is a natural state so deeply ingrained in historical and national narratives that it is taken for granted and invisible. Here the concern is with the way in which rootedness and movement have moral force.

Malkki's fieldwork among Hutu refugees in Tanzania documents these processes at work. She draws attention to how the transnational displacements undergone by these refugees is incorporated into or enables a series of narratives about nationality and identity. While one might expect that refugee status is, in Erving Goffman's terms (1963), a case of "spoiled identity," she finds that it is rather a source of categorical purity; being a Hutu refugee in Tanzania marks one as more distinctly Hutu, given both a disconnection from Burundi and one's inherently temporary status in Tanzania. Where Hutu ethnic identity had previously sat awkwardly alongside Burundi nationality, now it could be more vigorously and unproblematically asserted. As she notes, Burundi was a "mere state" whereas the imagined Hutu nation is a "moral community" formed in exile. Nevertheless, this vigorous assertion of pure Hutu identity is largely a feature of those she studied living in refugee camps as opposed to those who have settled in towns in Tanzania, whose status in Tanzania is no longer so temporary and who do not live liminally disconnected from Burundi and Tanzania. This is not to imply that they do not anticipate a return "home" and did not think of themselves as different from the Tanzanians among whom they lived. Rather, their imagination of their position is more cosmopolitan than nationalized, and they talk of their status and home in spatial rather than moral terms.

Like Munn, Malkki points to the ways in which spatial arrangements, presence, movement, and habitation have moral and cultural significance. These authors' focus is on the (user) experience of space. Space emerges as a relational, cultural object, and much of this cultural meaning—rootedness, morality, kinship, and responsibility—cannot be reduced to Cartesian coordinates or global positioning system (GPS) references.

This implies that technologies that seek to enhance, incorporate, or respond to the user experience of space may be limited by the representational schemes by which we are used to operationalizing it. If the user experience of space is cultural rather than cartographic, then an alternative foundation for design presents itself. This is a connection to design practice that goes beyond the traditional formulation of requirements but that can be entirely actionable. For example, these kinds of considerations led us to reconsider the role of spatial experience in the design of the Undersound System. Undersound focuses on the collective production of spatial experience through the patterns of movement and migration that people engage in daily (Brewer et al. 2007). It allows riders of a public transit system, particularly the London Underground, to share music via mobile phones. Music is tagged according to the places it has been, where it has entered the system, and where it has traveled. With an emphasis on locally produced music, the system strives to reflect the ethnic, demographic, and social diversity of the regions covered by the underground system in the music that flows through, providing a link between the underground and the surface, but also supplying a means by which the pattern of flows and movements of people through the space can be uncovered. Spatial structure in Undersound is an emergent property arising out of the interactions of people and objects. It is diverse, relational, actively produced, collective, dynamic, and non-Cartesian. At the same time, it is rooted in an experience of space that is fundamentally aesthetic rather than instrumental (Brewer, Mainwaring, and Dourish 2008). The system's goal is to reflect exactly this contingent and collective experience of space—one that shapes as well as is shaped by senses of collective identity and participation in ways significantly inflected by the ethnographic work such as that discussed above. What other technological and information systems could be designed within this set of frameworks, or others suggested by Munn and Malkki?

Toward a Generative Account of Ubicomp

As ethnographic accounts produced outside the domain of technology development, the work of Abu-Lughod, Lutz, Munn, and Malkki certainly does not present implications for design in the form in which they are often requested within ubicomp research contexts—a delimited set of short-term requirements or constraints on the design of contemporary or shortly anticipated technologies. As detailed and rich accounts of aspects of human experience that reach well beyond the specific sites at which research engagements typically take place, they certainly *do* present implications for design in the form of consequential, profound, and direct guidance for how to think about the issues in projects such as Undersound. Information technology and interactive systems are not in evidence in any of their studies; (user) experience, however, is front and center. This (user) experience is their topic, and to the extent that what they attend to is the role of emotion and mobility in (user) experience, their implications for the design of technologies in these areas are legion.

As we have argued throughout this chapter, the implications for technological design found in these ethnographic accounts are not of the requirements-capture variety. They set constraints on design, certainly, but not in terms of operationalizable parameters or specific design-space guidance. They may speak more to the way that information technologies and digital media take their place among the other "mediascapes" that shape cultural experience (Appadurai 1996). This is not a move away from design, however. What these studies in fact tend to do is open up the design space rather than close it down, talking more to the *role* of design and technology than to its shape. Implications for design encompass not only specific technological imperatives but also implications for how we go about design in the first place, what it might do, how it can be carried out, and what import it holds.

Importantly, these ethnographic understandings are derived not from the empirical aspects of ethnographic work; they come from its analytic aspects. That is, the ethnographic engagement is not one that figures people as potential users of technology and looks to uncover facts about them that might be useful to technologists (or marketers). Ethnographic engagements with topics, people, and field sites instead are used to understand phenomena of significance to design, and the implications arise out of the analysis of these materials. This goes again to the marginalization of theory that we discussed earlier in this chapter, in which the very fact that ethnography is conducted under particular analytic auspices is neglected or ignored.

It is also worth noting the temporal context and lifetime of ethnographic accounts, which can remain relevant far beyond their moment of writing. The studies we have cited were published between the late 1980s and 2000. We have cited others of relevance from earlier decades. Even if these studies had been conducted under technological auspices and had addressed design considerations, one has to ask what the implications for design would have been in 1995, 1985, or 1975, and what they would mean today. We have a feeling that they would tell us little about iPods, mobile phones, or blogs. Yet the theoretical contributions that the studies provide have a considerably longer shelf life and a relevance that transcends specific technological moments.

Our assertion is certainly not that design recommendations are poor things to include in ethnographies. Tight couplings of ethnographic materials and design practice have been both successful in design terms and productive for the research community—for instance, the Lancaster work on air traffic control (Hughes, Randall, and Shapiro 1993). The presence or import of implications for design, however, are not the only appropriate criteria by which ethnographic contributions can be judged. In fact, even in cases where such recommendations can be concisely and effectively formulated, to focus on those as the outcomes of ethnography at best distracts from, and often completely obscures, the analytic and conceptual work that lies behind them, which is frequently where the substantive intellectual achievement is to be found. What matters is not simply what those implications are; what matters is *why*, how they were arrived at, what kinds of intellectual (as well as moral and political) commitments they embody, and what kinds of models they reflect (Flyvbjerg 2001).

In thinking about ethnography (or indeed any social science contribution), it is crucial to distinguish two levels and two sorts of contributions: the empirical and the analytic. The empirical materials make up the fundamental observational material—the "this is what happens" detail of ethnography. The analytic materials comprise the ways in which these data are theorized, understood, organized, juxtaposed, interpreted, and presented in order to make an argument that reveals something about the setting under investigation. Observations are always theory laden, and any encounter between ethnographer and field involves a whole host of analytic positions, so a hard-and-fast separation would be impossible. But at least as far as ubicomp is concerned, we can distinguish between these two as ways in which an ethnography make a contribution—in terms of what it says happens and the ideas it offers for thinking about social life.

We would argue that the call for implications for design, drawing on the notion of requirements in traditional software engineering, is a request for empiricism. It is a request that the ethnography provide "facts"—when people work, how they talk to each other, what they do when they sit down at the computer, and so forth—which can be translated into technological constraints and opportunities. Certainly, many ethnographic studies can offer such things (although it is important not to ignore the role of the ethnographer as interpreter and framer of these "facts," rather than as a passive mirror of the site).

What has traditionally been more complicated has been to establish a deeper, more foundational connection between ethnography and design—to look for a link at an analytic level versus simply an empirical one (Dourish and Button 1998). The analytic contributions tend not to be seen as holding implications in the same way.

It is not that these do not have profound implications for design, because as we have seen, they do—indeed, often more profound than a laundry list of facts and features. Their impact, though, is frequently more diffuse. They provide us with new ways of imagining the relationship between people and technology. They give us ways of approaching design. Still, they typically go beyond specific instances of design. More to the point, they draw in general on the fundamental repudiation of a traditional separation between designer and user, between technology and practice. To the extent that these implications are not formulated as implications for design, it is because the categories of design, user, and designer are themselves in question.

Is it a cop-out to say that what these studies offer is a new framing for the questions rather than a specific set of design guidelines? Hardly. One obvious point to maintain about these reframings is that they have both a broader scope and longer-term impact than a simple series of requirements. They reach beyond the level of specific investigations. Is it a lack of imagination to fail to discuss technical matters? Again, hardly. What we have presented here are in fact acts of reimagining. In the cases that we have provided, technology was simply never in question in the first place, so naturally it did not feature as part of the discussion. But more generally, we are arguing that the movement from ethnographic engagement to design practice is inherently a conceptual and imaginative move, not a rote translation of empirical evidence into designed fact. Is it a lack of courage to claim that ethnographers need not provide implications for design? Yet again, hardly. If the push back is anything to go by, it takes considerably more fortitude to argue against the hegemony of design practice rather than to submit to it.

Perhaps it is a question of modesty. The engagement between ethnography and design must be just that—an engagement. Ethnography and ethnographic results are part of that engagement. The scope of the project of ethnographically grounded design goes beyond either ethnographic inquiry or design practice. Attempts to use ethnographic inquiry as a simple substitute for engagement with users, supplying a convenient summary of people's needs, goals, and meanings, are attempts to decouple design practice from its consumers and users. It is precisely these engagements that ethnographers seek to stage and frame in ubicomp and beyond.

The question is likewise one of responsibilities. We can ask this question in two ways. Whose responsibility is it to connect ethnographic results to design practice? According to the implications-for-design position, it is the ethnographers' responsibility. If the technological design implications are not clear, it is from the ethnographer's failure in meeting responsibilities. Certainly much could be said about the ways in which ethnographers need to frame results for broader publics (a concern that ethnography has long recognized), but we would contend that it is no more the ethnographer's responsibility to speak to design within the context of each specific publication than it is the designer's responsibility to speak likewise to ethnography. Rather, the responsibility for ethnographically grounded design results is a collective one.

The other way to ask this question is, To whom do ethnographers owe their responsibilities? Again, based on the implications-for-design position, ethnographers own their responsibilities to the design subcommunity. The alternative is that ethnographers owe their responsibility to their participants and informants, to the people with whom they have engaged, whom they represent, and for whom they speak (Rogers 1997). At times, that responsibility may be best served by engaging in technological or information systems design. At other times, it may be best served by heading off fruitless design activities, and, at still other times, it may be best served by reframing the questions. If the role of the ethnographer in ubicomp is to stage encounters between sites and technologies, then the forms that will take may vary considerably.

Perhaps one way of reconsidering the role of ethnography in ubicomp design is to question the concept of the ethnographic site. One of the more significant transformations of contemporary anthropological ethnography has been the concept of multi-sited ethnography, as developed particularly by George Marcus (1995). Whereas traditional ethnographies since Malinowski have focused on a geographically bounded field site, Marcus observes that in the context of globalization, culture can no longer be adequately circumscribed in such a manner. The Trobriand Islands can no longer (if they ever could) be approached as a "realm apart," but must be understood within a broader web of relationships to other parts of the world and other forms of cultural practice, including, for example, their connection to international academic anthropology and the cultural settings in which ethnographic results are presented (Marcus and Fischer 1986). Contemporary ethnography must concern itself instead with transnational flows of people, capital, and culture. This is perhaps especially relevant when considering information technologies—technologies that are both means and embodiments of these globalized practices. Miller and Slater's Trinidadian Internet, for instance, is a means by which cultural practice operates within a globalized economy. When we attempt to discharge the mythology of the field as part of a professional rite of passage, we are forced to consider the concept more critically (Gupta and Ferguson 1997).

What might happen if we started to think more critically about the site of ethnographic studies in ubicomp and HCI? In what ways can we separate the technical practices of one organization or set of users from those others with whom they interact, from whom they learn, and with whom they exchange information, artifacts, and people? We might, say, reconfigure the ethnographic project in ubicomp by thinking of studies not as independent investigations but rather as contributions to a broader ethnography corpus whose site is not a particular office, campus, or city within which technology is used, but rather it is the global technology culture itself, or the intersection between cultures of technology production and consumption. Certainly, this suggests that we might need some very different criteria for assessing the role and contributions of ethnographic studies within ubicomp.

In this chapter and throughout this section, we have articulated a different kind of ubicomp—one that is a form of social and cultural production as much as a scientific research trajectory. We have been particularly interested in mapping out a new set of relationships between ethnography and ubicomp that move beyond formulaic implications for design to something considerably more generative and inviting. In the next chapter, we take this one step further and offer a critical interrogation of contemporary ubicomp preoccupations.

II

By many accounts, ubicomp has been tremendously successful. It has been successful as a research endeavor. In addition to being a topic in its own right, it is also a central aspect of the research agenda of many other areas of computer science research, from theory to embedded systems. Furthermore, it has been successful as a technological agenda, meaning that Weiser's model of a single person making use of tens or hundreds of embedded devices networked together is a reality for many people. Throughout this book, however, we are interested in posing the question of the relationship between these two successes.

The foundational elements of the ubicomp vision—a future in which our encounters with the world and each other are smoothed by the application of technology, a world to be delivered to us by heroic engineering—is remarkably persistent. There is a fundamental technological determinism at work, and Peter Tolmie and his colleagues (2001) note the irony of, on the one hand, Weiser's attempt to move beyond the idea of the "dramatic computer" and, on the other, an inevitable research practice celebrating ingenious design. Weiser and other early ubicomp visionaries provided an imaginative vehicle for understanding the encounter between technology and the social world—one in which technology would play a liberating role. This is both an alluring vision and a common one (Kling 1994). Nevertheless, while the majority of ubicomp's research attention has traditionally been devoted to the proximate future, just around the corner, we have suggested instead that some twenty years after Weiser originally formulated it, ubiquitous computing has indeed arrived. If the availability of devices with wireless data communications and powerful computational properties is anything to go by, then it is hard to deny that computation is already ubiquitous. This raises two interesting questions: First, why did we fail to notice it? And second, what should we do as a consequence?

As we have written throughout part I, our failure to notice the arrival of ubicomp is rooted (at least in part) in the idea of seamless interoperation and homogeneity. The ubicomp world was meant to be clean and orderly; it turns out instead to be a messy one. Rather than being invisible or unobtrusive, ubicomp devices are highly present, visible, and branded, but perhaps still unremarkable in the sense explored by Tolmie and his colleagues. Ubicomp has turned out to be characterized by improvisation and appropriation; technologies lashed together and maintained in synch only through considerable efforts; surprising appropriations of technology for purposes never imagined by their inventors and often radically opposed to them; widely different social, cultural, and legislative interpretations of the goals of technology; and flex, slop, and play.

The material in this section grows out of a series of papers that we have produced over the last several years, which were inspired by our collaborations with colleagues, topics of immediate concern, and dialogues (both formal and informal) at research conferences and in the published literature. The topics that we have presented here as sites in which to investigate the technological imagination of ubicomp—infrastructure, mobility, privacy, and domesticity—are scarcely the only ones we could have selected or even, arguably, the most important ones; they just happen to have been the topics that preoccupied us at one moment or another. We have said next to nothing here of religion and spirituality, romance and intimacy, health and well-being, birth and death, citizenship and civic participation, or work, leisure, and sport—all topics of crucial significance to people every day; what is more, all topics in which digital technologies of the sort we have been discussing play ever more central roles. Although we have neglected these topics here, no significance or symbolism is implied; equally fruitful investigations could certainly be conducted at those sites.

We are perhaps less concerned with the specific sites or topics of investigation than we are with the styles of investigation as well as the forms of scholarship that we have been trying to develop, exemplify, and advocate here. It is, for us, an approach entirely in keeping in a number of ways with the constitution of the ubicomp program as originally set out by Weiser. First, it attempts to take seriously what it might be to make a claim for a ubicomp, not in the sense of subscribing to that as a plausible goal but in understanding the ways that computing is both embedded within and constitutive of a world of shreds and patches, characterized by difference, disjunctions, and distinctions. The idea and practice of ubiquity—the work that needs to be done, by designers, users, legislators, and operators, to produce specific and limited forms of ubiquity—are themselves useful topics of interest, as are the areas of life, peoples, practices, and worldviews excluded from this apparently expansive perspective. Taking ubicomp seriously thus implies taking seriously the practices of ubiquity. Second, our approach draws on the broad intellectual vision that Weiser similarly sets out as a component of the ubicomp agenda from the outset. From the earliest accounts, ubicomp is no narrow topic located entirely within the boundaries of a traditional computer science. It expands those boundaries and reaches beyond them. As we argued earlier, we feel that this calls for integrative rather than parallel investigation from different disciplinary perspectives—another form of ubiquity perhaps, and one that we acknowledge as no less problematic.

It is precisely these problems that make the area of ubicomp so productive. We suggested in chapter 2 that what characterizes ubicomp settings is their messiness, and as we have proceeded through other topics, we have found this messiness to pervade methods, concepts, and practice. Yet we are not implying that the messiness is an obstacle, nor are we proposing that it would be appropriate to tidy up. We find messiness inspiring, productive, generative, and engaging. Tidiness is static, rigid, fixed, and closed; messiness is dynamic, adaptive, fluid, and open. Likewise, when we talk about the problems of ubiquity and the impossibilities of erasing difference, we do not see these as nails in ubicomp's coffin but rather as points of interest and exploration. Cleaning things up has, in short, neither been our goal here nor has it become our goal going forward; embracing the messiness gives us so much more to think about.

In this section, we turn to contemporary ubicomp practice. We do this by exploring a set of related yet distinct themes that arise both in the research literature and as aspects of everyday experience. Given the multifold direct and indirect roles of digital technology in contemporary life along with the expansive nature of the ubicomp vision, it is perhaps unsurprising that these different themes are not easily untangled. Nonetheless, each topic provides us with an entry point and a particular set of phenomena that help to illuminate different aspects of the overall story.

Each of the four topics that anchor this section's chapters—infrastructure, mobility, privacy, and domesticity—appear at first glance reasonably contained. Infrastructure, it would seem, denotes a specific part of the fabric of ubicomp technology; mobility is a limited concern that places technical demands on some subset of applications; privacy, while a pervasive consideration, is one that arises only at the point of technology use, not technology design; and domesticity is a consideration that emerges in bounded application domains. In selecting these as our points of entry into what we might more broadly call the "ubicomp imagination" or mythology, however, we deliberately resist the constrictions of these apparent boundaries. Indeed, what is most interesting about these topics—and the reason that we focus on them—is exactly the way in which they rapidly expand beyond any natural boundary, and we can find, refracted through them, much more expansive aspects of the broad ubicomp vision. Infrastructure enlarges to incorporate questions of standardization, compatibility, interchange, uniformity, and the work involved to create local as well as temporary stabilities within technological and social environments continually in flux. Mobility raises questions of stability too, although here in terms of space, scale, the cultural significance of patterns of movement and habitation, and in turn the commitments that ubicomp makes to the relationships between place and social life. Privacy, while a concern that pervades much public and scientific discourse about technology and everyday life, dissolves as a concept and gives rise instead to a broader series of considerations about conceptions of information as an object of exchange along with the social conventions that surround our orientations toward risk, secrecy, and identity. Finally, domesticity marks out not simply a particular region for technology deployment but also a set of historically, geographically, and culturally grounded commitments to the nature of social relations much more broadly. Each of these therefore provides a lens through which we can examine an aspect of ubicomp's imagination about the relationship between technological and social life.

Each of these topics is an opportunity to engage in the sort of examination that we advocated toward the end of chapter 4—one that is grounded in empirical engagement with everyday life but that seeks within that an account of the patterns through which we construct coherence in the course of daily practice, with, through, and around technologies of all sorts, digital and otherwise. What they reveal is that ubicomp offers even more as a site of examination than simply an account of contemporary, cutting-edge technology development; it allows for an ongoing working out of the nature of technologically mediated society.

5 What Lies Beneath

All computational developments are in some sense infrastructural in that what computer designers, software developers, and information systems architects attempt to do is to create platforms for people to achieve their own individual goals. Like the familiar infrastructures of the built environment, from highways to the electricity grid, computational systems exist not for their own ends but rather to be used to some other effect. Arguably, then, to become infrastructure is a mark of a successful technology—it becomes unremarkable. This was certainly one of Weiser's explicit goals in his original statement of ubicomp's vision. His opening remarks speak to the nature of infrastructure: "The most profound technologies are those that disappear. They weave themselves into the fabric of everyday life until they are indistinguishable from it" (Weiser 1991, 94). This is not the "disappearance" of rapidly miniaturizing technology that becomes so small it literally shrinks from view; instead, this is an infrastructural disappearance, a form of easy habituation and invisibility in use that follows from casual familiarity. This sort of infrastructure is still quite visible if you go looking for it; it is just that, in most cases, we don't.

In this chapter, we examine ubicomp through the lens of infrastructure. Given that infrastructure is something typically unseen and unspoken, it may not seem like a fruitful topic for social or cultural inquiry. Yet as Star (1999) demonstrates, in a paper that she delightfully dedicates to fellow members of the "Society of People Interested in Boring Things," infrastructure is a potent topic for ethnographic investigation, encompassing the processes by which infrastructure standards are developed and deployed, the dynamics of infrastructure development and maintenance, the constraints that infrastructures impose, and the forms of erasure and homogenization in which they engage. It is through a closer examination of infrastructure that we can return to the messiness we first spelled out in our introduction.

Infrastructure?

Because "infrastructure" is the key term around which our explorations in this chapter revolve, we should begin by unpacking what we want it to do for us. By infrastructure here we are not simply referring to the technological substrate of networked services that support the development and deployment of pervasive computing applications (although those are certainly relevant to our argument). Rather, we will take the term more broadly as well as literally as pointing to the structures that lie below or beneath the surface of applications and interactions.

Infrastructure is normally taken for granted, almost by definition. Star (1999), however, has spotlighted the value of infrastructure as a topic of ethnographic inquiry, both as a means of uncovering the unspoken conventions of everyday practice and a way of unpacking the implicit relationships between different communities, interest groups, and perspectives. In this light, infrastructures operate as another manifestation of cultural practice and can be scrutinized as such. Star points to nine properties of infrastructure: embeddedness (it is "sunk into" other structures, social arrangements, and technologies); transparency in use; reach or scope (going beyond a single event or site); learned as a part of membership in a community of practice; linked to conventions of practice; embodied as standards; dependent on an installed base; visible on breakdown; and fixed incrementally rather than globally. For the purposes of this chapter, there are at least two perspectives on infrastructure that are relevant here.

The first is a sociopolitical reading of infrastructure, from which we might examine infrastructures as crystallizations of institutional relations. Infrastructures drive and maintain standardization, reflect and embody historical concentrations of power and control, and are instruments through which access is managed. As a number of commentators have observed (e.g., Castells 2000; Harvey 2001), despite the revolutionary and transformational rhetoric surrounding the development of networked information infrastructures, in practice they are as likely to reinforce as to destabilize existing institutional arrangements. For all the contemporary interest in blogs, user-generated content, and individual publishing, for example, information on the Internet tends to be centralized in largely the same hands as that in other media. From a sociopolitical perspective, then, we might be concerned with the governance and regulation of the wireless spectrum, the forms of control embodied in the "common carrier" and "bilateral-peering" arrangements through which telecoms and Internet service providers manage flows of traffic, or the use of firewalls and related flow-control technologies to transform, shape, and manage the virtual network over which pervasive services are deployed. At the time of writing (in early 2009), two cases seem pertinent. In the first, the experience of the Internet's use in countries such as China, Turkey, and Singapore, where information flows are subject to considerable state restriction, put lie to the idea that "information wants to be free" (Turner 2006); technological arrangements and commercial interests can allow exactly the same sorts of restriction and management that are deployed around "old media." The second example relates to ongoing debates about what is termed "net neutrality"—a political and regulatory debate concerning whether ISPs should be able to introduce traffic-routing policies that favor particular sorts of network traffic (such as that to and from their commercial partners).

Economics plays a role on a smaller scale too, of course, when we bear in mind that the infrastructures in which we are interested are largely ones deployed and operated by commercial entities operating in a complex marketplace. Mobile telephony serves as an example once more, where the impact of calling plans and pricing policies should not be underestimated when thinking about what phones do and how they do it (Palen, Salzman, and Youngs 2000). In the European Union and most of India and China, consumers do not pay for incoming calls. In India during the first years of this decade, although Bombay, Bangalore, and Calcutta did not charge for incoming calls, Delhi's local mobile companies did; as a result, there was a marked absence of mobile phone use as compared to the other cities. In London, at least, more than 50 percent of children aged fourteen to eighteen have mobile phones, in no small part because the various service providers have developed pay-as-you-go plans that do not require binding legal contracts, because it is legally impossible to negotiate with minors.

From this sociopolitical stance, ubicomp is just one of a number of recent infrastructural developments. Stephen Graham and Simon Marvin (2001) have pointed toward a number of trends, including the growing dependence on infrastructures for everyday life, increasingly contested forms of interoperation and standardization, and an ever more complex regulatory environment within which these issues are embedded. This, they argue, is leading to an increasingly fragmented and splintered experience of urban space; indeed, on a more local scale, the everyday users' simultaneous juggling of many forms of network infrastructure (i.e., wired Internet, wireless Internet, dial-up, GSM, code division for multiple access [CDMA], etc.) may be an example of their thesis at work in the information technology domain. The spaces through which we move become visible in terms of their network accessibility and, consequently, in terms of their implied electronic "locality."

As such, this sociopolitical account is one way that we might examine the relevance of infrastructure to ubicomp, and indeed the sociopolitical context is our focus in chapter 2. The second approach is an experiential reading of infrastructure, which concentrates not so much on how infrastructures reflect institutional relationships but more on how they shape individual actions and experience. It is this perspective that primarily concerns us here, in two ways: what we refer to as "the experience of infrastructure" and "the infrastructure of experience."

By the experience of infrastructure, we point to the ways in which infrastructure, rather than being hidden from view, becomes visible through our increasing dependence on it for the practices of everyday life. By the infrastructure of experience, we draw attention to the ways in which, in turn, the embedding of a range of infrastructures into everyday space shapes our experience of that space even as it provides a framework through which our encounters with space take on meaning. The experiential reading of infrastructure, then, sees infrastructure and daily life as coextensive; accordingly, it encompasses not just technological but also the social and cultural structures of experience in ubicomp settings.

Space and Infrastructure

As we noted above, we take a broad perspective on infrastructure, to better examine the kinds of considerations that animate Star's analysis. To do this, we focus on space as infrastructure. We make this choice partly to defamiliarize the notion of infrastructure by avoiding the more obvious sorts of infrastructure with which ubicomp is plainly concerned (such as electric power, software architectures, and hardware platforms), and partly because from this stance the relationship between interactional and technological arrangements can be seen a little more clearly. From the sociopolitical perspective, infrastructure is a natural topic for discussions of space along with the distribution of activities, power, and movement. When we think about the experience of space, though, we see that infrastructures play multiple roles. Streets provide an infrastructure for the movement of people and goods about a city. Yet the naming of streets is simultaneously an infrastructure for encountering and experiencing the city in terms of regions, paths, and flows—street naming defines patterns of sameness and difference that critically define what you see when you look around. Of course, some urban areas never name their streets at all but instead rely on a set of sociospatial directions to guide an individual or mark a journey. Certain cities, in this way, become untraversable to those not already resident within them—the location markers are not abstract demonstrations of the city but rather concrete manifestations of social relationships, historical events, and institutional memories. Pune, India, is one such city; when Bell did fieldwork there in 2001, all directions and addresses were given in reference to key landmarks and buildings. More complicated still, the names used were often those for previous owners, institutions, and occupants. This kind of remembered spatiality would not find a comfortable home within, say, the addressing practices of IPv6—the newest Internet addressing protocol.

In computer science, both at a foundational level and in terms of the design of applications, Western conceptions of spatiality have played a central role in how computation is conceived, modeled, and presented. Phrases like "the information superhighway," "World Wide Web," "ad hoc network," "virtual town hall," and even "surfing the web" all encode specific forms of spatial practice and knowledge. In collaborative systems, the most extreme example is perhaps the development of collaborative virtual environments in which virtual worlds provide a setting for the action and interaction of "embodied" characters (e.g., Benford and Fahlen 1993; Churchill, Snowdon, and Munro 2001). Even absent such literal interpretations of the role of spatiality in everyday experience, however, spatial metaphors of computational phenomena—be those name spaces, work spaces, file spaces, shared spaces, web spaces, and so on—are persistent features of computational practice.

Yet despite the relatively widespread appeal to scientific and mathematical accounts of spatiality as a foundation for interaction with computational phenomena, space itself remains relatively unexamined in computer system design. If time, as John Archibald Wheeler commented, is what prevents everything from happening at once, spatial metaphors and models in interactive system design are invoked mainly to allow for computational objects to be kept apart from each other. Separation allows distinctions to be drawn. Files in a file space can be distinguished from each other and clustered according to needs; activities in a work space can be kept from interfering with each other; conversations between different actors in a collaborative virtual environment can proceed independently.

In contrast to this instrumental model of space, we consider spaces as infrastructures, and in particular as layers of infrastructures, human and technological (McCullough 2004; Klein 1997). We refer not only to physical infrastructures but more broadly to infrastructures as fundamental elements of how we encounter spaces—infrastructures of naming, mobility, separation, interaction, and so on. In so doing, we are foregrounding an interest in the cultural constructions of space and in turn infrastructure. Infrastructure is analytically useful because it is both embedded in social structures and serves as a structuring mechanism in itself. It is this dual role that especially interests us here—how the infrastructures of space and ubicomp are mutually, reciprocally coupled to social and cultural practices.

Encounters with many different infrastructures shape the experience of space. Transportation systems are an obvious instance. For example, when first visiting Paris and traveling on the Metro, one experiences the city as a series of islands connected by Metro stops—until one day one walks down the street, realizes that some of those stops are only a few blocks apart, and starts to experience the city as a continuous phenomenon. At the same time, iconic representations of transportation infrastructures themselves become mental models of spatial organization (Vertesi 2008). Religious sites or institutions (i.e., churches, temples, and mosques) suggest a different sort of urban infrastructure, not merely as destinations in and of themselves—fixed points on a specific sort of encounter within a city as resident, tourist, or pilgrim—but also as manifestations of inter- and intraurban connections. Schoolchildren in Britain, and indeed all over the former British Commonwealth, grow up with mnemonics to remember the various sounds of London's churches ("Oranges and lemons, say the bells of Saint Clement's")—a city's soundscape reflected as nursery rhyme so one never gets lost (Garrioch 2003). In contradistinction, mosques all over the world orient themselves to Mecca—Islam's holiest city—suggesting a different kind of invisible geography or infrastructure rarely accounted for in current theorizing about the city or the mobile technologies therein. Traffic flows, service times, calls to prayer, regions, and neighborhoods are all infrastructures that shape one's experience by making it meaningful in different ways; in turn they are shaped and configured in support of patterns of social practice.

The Practical Organization of Space

In focusing on the practical organization of space, we want to emphasize the mutually constitutive relationship between collective understandings of space and the practices and activities that people carry out in them. Here we cast our net broadly, surveying studies of work, transportation, leisure, and domesticity. In all these domains, we argue, there are complex and dynamic interplays between the physical dimensions of space and the social actors who inhabit those spaces, however temporarily.

Spaces have structure and meaning for us in terms of our relationship to a variety of systems of practical action and interpretation. In studying conversational practice, Emanuel Schegloff (1972) notes the range of ways in which place is "formulated" in conversation—that is, how a particular formulation of location among an almost-uncountable set of alternatives is selected and used in the course of spoken interaction. The interactional determination of an adequate formulation, he shows, is much more than simply a selection from a hierarchy of degrees of ambiguity; the use of an appropriate formulation both depends on, and displays the use of, collectively shared understandings of practice. The question "Where are you?" might have any number of answers, not all of them relating to place, and be contingent on the work at hand; the determination of an appropriate response, as a part of an ongoing interaction, is one aspect of competent practice, whether that practice is astrophysics, plumbing, or passing the time with a stranger at a bus stop. Space and practice are similarly tied together through the processes of interpretation that accompany embodied practice.

Charles Goodwin (1994, 1995) uses the term "professional vision" to talk about the specialized visual and discursive practices associated with disciplines such as archaeology and oceanography, in which acts of seeing arise out of theoretically structured encounters between people, instruments, and the world. He contends that "the ability to see a meaningful event is not a transparent psychological process but instead a socially situated activity accomplished through the deployment of a range of historically constituted discursive practices" (Goodwin 1994, 606). The ability to see a place—and see it *as* a place and as a *particular sort* of place—depends on similar practices.

While the studies of the practical organization of space focus in detail on highly localized settings, the issues of the relationship between space, experience, and practice play out on much broader scales. Here, by looking across contexts, we are concerned with people's experience of spaces and landscapes as cultural, not least because it is this experience that is disrupted and transformed when new technological opportunities enter those spaces.

In a series of studies, Christian Heath, Paul Luff, and their colleagues at King's College London have applied principles and methods from conversation analysis to scrutinize and unpack the mutual coordination of actions in a variety of settings of collective activity, ranging from doctors' offices to transportation control centers to art galleries. Their study of London Underground control room operators uncovers the delicate coordination between the activities of team members who have different responsibilities but must nonetheless coordinate their actions to achieve a coordinated effect (Heath and Luff 1992). For instance, they show that control room staff attend not only to their own working activities but also to those of others in their periphery so as to anticipate upcoming action and coordinate their own alongside it. The controllers go so far as to organize their activities in such a way as to allow others in the immediate local environment to observe and interpret it as an aid to this process. Again, although the work of the control room staff is organized in terms of separate responsibilities, formal processes, and information flows, in practice the fact that it arises in real time in a common space provides the participants with the means to coordinate actions in a much more integrated way. Katrina Jungnickel (2004) documents a similar coordination of work practices between driver and conductor on the now-defunct Routemaster bus plying the number 73 line in London. Although conductors communicated with drivers solely via a one-tone bell, there were distinctive patterns between particular drivers and conductors. What becomes clear here is that work practices are not only embedded within and constrained by the specificities of particular spaces (a moving bus, a control room, or a doctor's office) but they are also learned, familiar, and relational.

The infrastructure of work spaces are not the only ones that are negotiated through learned, familiar, or relational practices. The "home" is also a site of such negotiations, a space with its own distinctive infrastructures and understandings thereof, and while we will discuss domesticity at length in chapter 8, the infrastructures of home spaces are worth exploring here. There is a large body of ethnographic and sociological work centered on the home as a space of daily activity; attention is paid to notions of public and private within the home, the gendering of different domains within the house (kitchen, shed, living room, and home office), informal divisions of labor (Strausser 1982; Livingstone 1992) for household tasks, and the physical divisions of space. Much recent work on the home has traced the impact of the introduction of new technologies on the rhythms and rituals of such spaces within a Western context (e.g., Lally 2002; Silverstone and Hirsh 1992; Rodden et al. 2004; Livingstone 2002). But here too, cultural practice and geopolitical institutions have a significant role. Homes outside the West often exhibit different infrastructures and practices. Not only are they embedded within fundamentally different systems of meaning but they also exist within different sorts of physical, infrastructural, and legislative contexts.

Bell's multi-sited ethnographic research in urban settings in six Asian countries in the 1990s explores these different kinds of models of domesticity, infrastructure, and space (Bell 2005, 2006b; Bell, Blythe, and Sengers 2005). Unlike their U.S. counterparts, for example, not only are urban Asian homes generally smaller, with fewer rooms, they are rarely freestanding dwellings; homes are far more likely to be apartments within larger buildings or complexes. Electricity, water, and various forms of communication infrastructures are neither evenly nor consistently distributed. In these dwellings, residents might share resources, including common areas and infrastructure. Resources, for most urban households, are distributed at a building rather than an individual space level. In China, cable services are provided to the apartment complex and individual households receive only what is bundled to the building complex; there is not a model of individual customized subscription. Yet resource allocation certainly follows certain social/political hierarchies, and there are more channels available in apartment complexes that house upper-middle-class families and business owners than those that house factory workers. It is possible to imagine that such distribution and control of domestic infrastructures might in turn generate practices of resource management within as well as around the home that are not well supported in current visions of a computationally augmented living space. Crosscutting the infrastructures' fixity in particular places are also a set of infrastructures that support mobility, as we discuss in chapter 6.

Related research has shown the same issues in a leisure setting: the art gallery (vom Lehn, Hindmarsh, and Heath 2001). Art is not experienced statically and, critically, in isolation but rather in a space that is moved through and occupied simultaneously by others, both companions and strangers. While the experience of art is often a private phenomenon, it frequently transpires in public spaces. The presence and activities of others configures the space for a gallerygoer, directing attention, constraining and guiding movement, and so forth. Similarly, the ways that exhibits are encountered sequentially as the gallerygoer moves through the space, places the individual pieces within a broader experiential context; they may be encountered not so much individually but instead collectively, as configured by conventional patterns of movement. Particularly in the case of exhibits in science museums (rather, perhaps, than traditional art gallery spaces), where the exhibits themselves may be interactive and require participation by the gallerygoers, people are seen to be highly responsive to the presence of others along with the public availability of interaction between people and exhibits. People dynamically construct collaborative encounters with the exhibits: "the visual, vocal, and tactile contact of others provides resources for looking, seeing, and experiencing the various exhibits" (ibid., 206).

These examples, drawing on widely different domains, highlight the complex relationship between the physical structure of space, the local organization of the activities within it, and the collective practices of those who occupy it. The choreography of mutually directed activities within space furnish it with a local logic that, in turn, makes those actions meaningful to the parties involved. Spaces are inhabited. Actions are not merely "played out" in space; they serve to structure and organize that space. The movements of people—the places they gather and those they avoid, the places they talk and the places they sleep—structure space; the logics of space are enacted in and through everyday life. So while much of the discussion of spatiality in interactive systems has conceptualized space as a passive "container" within which decontextualized actions may be arrayed, our infrastructural perspective has attempted to highlight the mutually constitutive nature of space and practice.

In contrast to iconic and frequently used tropes of urban spatial infrastructures, it is helpful to look to central Australian Aboriginal peoples' experience of the land that they occupy (Bell 1983; Povinelli 1993; Stanner 1958; Munn 1996). There have been interesting and innovative technology interventions into Aboriginal communities in central Australia, including early satellite and television deployments (Michaels 1986, 1987), but we are more interested in exploring the cultural construction of space along with the radically different imaginings of place, infrastructure, and mobility.

As we discussed in chapter 4, local indigenous belief systems charge the Warlpiri and Kaiditch (Kaytetye) peoples of central Australia with a ritual responsibility for the land. This reflects a symbolic dependence, not simply an ecological one; the people's responsibility is not just for environmental stewardship of the land but also for Dreaming it into existence. The existence and persistence of the land and the landscape are inextricably bound up with the people and their cultural practices. More significantly, this relationship means that the landscape is not simply a physical topology but also a cultural and historical one. This arises in a number of ways.

First, the visible features of the landscape are held to be the results of the action of creatures during the time of creation, often described as the Dreamtime, as mentioned earlier. The features of the natural world are the symbolic footprints of these metaphysical creatures, which resemble animals familiar from daily experience. These totemic creatures are associated, too, with tribal groups and lineages, creating a direct relationship between social groupings and the land. In every telling of the creation stories, in ceremony and beyond, the landscape and its importance is foreground, and so it plays a central role not merely in stories of the past but in the experience of the present as well. This historical connection is manifested in the sacred status of particular sites. By definition, knowledge of sacred sites is not public; it is restricted to those who have a responsibility for those sites. So, again, the landscape takes on a layer of meaning through these responsibilities and forms of local knowledge.

Second, according to their belief system, all meaningful actions and events leave their imprint in the land. As we have noted, the features of the landscape are associated with the actions of various creatures during the Dreamtime, but this link between space and action extends also to human activities. Battles, celebrations, births, deaths, and other events of human history also leave their resonance on the land; the landscape is simultaneously a physical and historical one. These historical resonances take on ritual significance, yet they also permeate the experience of everyday life, making daily life as well as the movement through space a cultural and historical experience. Stories become manifest in the landscape itself, and historical experience is rooted in physical space, which in turn becomes a way to maintain connection to the past and the events that shaped current experience.

Third, kinship groupings are also projected into the spatial domain. Kinship is a dominant aspect of daily life. For instance, since the names of the dead are not to be spoken under taboo, the dominant form of naming and everyday address is actually one based on kinship relations. These broad patterns are also tied to historical developments. All these influence the creation and experience of spatial relationships. The orientation of settlements reflects ancestral migration patterns, and their layout also mirrors local kinship and lineage relationships (Bell 1983). Areas of land or regions of a settlement are interpretable as associated with various groups, which themselves have significance in terms of kinship responsibilities, alliance and marriage opportunities, and so on. Kinship, then, provides a local logic for the experience not just of others but also of space itself; lineage patterns are inscribed into space. This may have the effect of limiting or transforming patterns of movement, as kinship patterns may make certain encounters inappropriate or uncomfortable (e.g., between people who stand in the actual or potential mother-in-law/son-in-law relationship); the ways in which kinship relationships and ritual exclusions are mapped onto the space becomes a primary way in which space is encountered, experienced, and navigated (Munn 1996).

This particular set of cultural arrangements elaborates on the view of space as a layering of infrastructures, background or invisible structurings that underpin mundane experience, and everyday action. The very organization of space—and then its use, occupancy, navigation, and so forth—are experienced through a range of cultural lenses that give it meaning and significance. Practice and action, too, take on meaning through the way in which they relate to these cultural scripts. The world of everyday experience is not simply the physical or visible world but rather one imbued with historical, social, and cultural meaning that is, critically, mapped onto and experienced through spatial patterns or, perhaps more accurately, habitation patterns. Everyday space is not experienced neutrally; it is experienced as inhabited, with all that entails.

The different spaces we have described over the last few pages— London Underground control stations, urban Asia, art galleries, and Aboriginal central Australia —have their own sorts of infrastructures—physical, social, historical, cultural, and political—forming complex layers and layerings of meaning, practice, and ritual. We argue that any form of pervasive computing deployment must not only contend with these layerings but indeed find ways to nestle between them. It is not possible to erase these layers with a Wi-Fi network or to negate centuries of history and cultural practice with a sensor deployment. Rather, we must pay attention to these infrastructures, reading their subtle and not-so subtle effects on the spaces and practices that we hope to augment with new technologies.

Regulating Infrastructure

When we think about infrastructure, we must also think about its regulation. Here we mean to cast regulation broadly to encompass both formal mechanisms as enacted through governmental, judicial, and economic institutions, and also more tacit forms of regulation via social and cultural institutions and practices. Formal mechanisms regulating infrastructures have impact on everything from appropriate forms and formats of content, pricing, decisions about component parts and standards, data traffic speeds, and backhaul rates. As we discuss in chapter 2, there are clearly ways in which governments can set terms for all manner of infrastructural deployment and use; Korea and Singapore are both strong examples of this. There are also less-explicit forms of regulation and participation, however. Tacit forms of regulation might include family choices about service providers and device placement within the home, religious proscriptions about device types and usages, and the placement of infrastructural nodes. That there is such a complexity of regulations also implicates a similar level of complexity when it comes to institutions (formal and otherwise) that imagine they are, or should be, stakeholders in infrastructural deployment and use.

In February 1840, the British colonial forces then present in New Zealand negotiated a treaty with the Maori—the islands' indigenous inhabitants. Although there have been serious debates about the nature of this negotiation and issues reconciling the two versions of the treaty, it has had lasting implications for landownership, resource management, and more recently, infrastructural build outs. Article II of the Treaty of Waitangi set forth that in exchange for protection from the British Crown, the Maori would be allowed to retain "their chieftainship over their lands, villages and all their treasures [*taonga*]." *Taonga*, in Maori culture, "refers to all dimensions of a tribal group's estate, material and non-material—heirlooms and *wahi tapu* [sacred places], ancestral lore and *whakapapa* [genealogies], etc." Over the subsequent 170 or so years, the courts and governmental authorities have interpreted Article II in particular to give Maori rights in fishing, mining, and forestry. In 2000, as the New Zealand government commenced the lengthy process to complete a 3G spectrum sale, lawyers acting on behalf of the Maori nation came forward with a novel interpretation of Article II. They argued that the air through which the spectrum passed was covered by this provision, it was a *taonga* and that, as such, the Maori should be granted rights to it, as they had in previous resource cases (Crehan 2000; Landzelius 2007). As a result of this legal challenge, the New Zealand government has set aside a certain percentage of the spectrum for Maori use.

Although there are few treaties as comprehensive as this, and few nations are covered by just a single foundational document, this example points the way to a much larger conversation about what it means to deploy infrastructure and how the spaces into which such infrastructures will achieve scale are understood. In New Zealand, indigenous religious practice, colonial encounters, and a 170-year-old document all figured into the conversation. What might this mean for other deployments? Who should participate in the discussion? Whose opinions and experiences are relevant? How might such individuals and institutions be included in both the conversations about deployment and those of regulation?

Sociality, Spatiality, and Ubicomp

What we are suggesting is an alternative model of space and spatiality than that which dominates current discourse in the design of ubicomp technologies and environments. Ubicomp brings computation out of the traditional desktop and into the spaces beyond; but the critical feature of these spaces is that they are always and already populated and inhabited. More to the point, the experience of space is that of multiple infrastructures—infrastructures of naming, movement, interaction, and so on—and these infrastructures emerge from and are sustained by the embodied practices of the people who populate and inhabit the spaces in question. Spaces are not neutral, and their complex interpretative structure will frame the encounter with ubicomp; by the same token, the opportunities afforded by new technologies allow for a reinterpretation and reencounter with the meaning of space for its inhabitants. The experience of space is fundamentally coextensive with the cultural practice of everyday life.

So how do these examples of the practical and cultural organization of space inform our understanding of the relationship between spatiality and emerging trends in pervasive computing? Here we again return to our interest in the physical manifestations of infrastructures and their cultural framings. We consider three topics: the physical reality of virtual environments, the spatial nature of mobile services, and the cultural framing of space.

The first of these is what we might term "the physicality of the virtual." In the age of networked environments, such as those of *Everquest*, *Ultima Online*, and *Second Life*, the rhetoric of virtuality suggests the electronic domain as one that exists apart from the everyday world (Miller and Slater 2000; Woolgar 2002; Boellstorff 2008). The technologies of ubicomp, however, require that we confront the physical reality of virtual environments. Wireless network infrastructures may be invisible and intangible, and they may facilitate participation in a virtual world, but they are themselves thoroughly physical. Anyone who has had to step outside a building in order to get a better cell phone signal or has wandered through a conference hall, laptop in hand, checking their Wi-Fi signal-strength meter, has firsthand experience of this overlay of the physical and virtual. The physical configuration of electronic services creates a new layer through which the physical environment is experienced and understood. Our physical encounters with space are mediated by the differential distribution of computational elements within that space, both those embedded in the fabric of the space itself and those carried by other occupants. Consider the ways in which wireless networking might be strategically placed to support informal encounters in a conference center or public space, or the difficulty that one might experience trying to walk through a crowded space without letting a Bluetooth cell phone come within beaconing radius of any other Bluetooth devices.

In some cases, this forces us to confront aspects of our development of urban space and the built environment. It is well known, for example, that GPS operates poorly or not at all in the urban "canyons" of New York City, where satellite line of sight is hard to achieve consistently. GPS, then, embodies a model of space and spatiality that provides a new lens through which to view the relationship between space and function. A similar set of observations has been made about mobile broadband in Sydney (Jungnickel and Bell 2008). Indeed, the physicality of the virtual exposes a range of cultural scripts and expectations that are embedded within a range of pervasive technologies. "Residential" access points for wireless Ethernet service are typically designed with a range of around 150–300 feet in three dimensions; they are designed, that is, for typical U.S. homes, but not for the high-density apartment living that characterizes many parts of Europe or Asia (Bell 2006b).

Inverting this relationship, we can also see how certain configurations of space make them particularly amenable to different forms of pervasive computing through the ways in which pervasive computing technologies interact with other spatial infrastructures. These examples all indicate that the virtual technologies that will ultimately facilitate pervasive computing have real, physical manifestations and impacts; wireless devices may connect wirelessly, but they do it through spaces occupied by buildings, people, and stories. Furthermore, they compete with other existing infrastructures of space, time, architecture, and even weather. It is also the case that different regulatory bodies, government structures and strictures, and cultural patterns can profoundly shape how the virtual is materialized.

A second aspect of this relationship between ubicomp and social understandings of space is the spatial situatedness of mobile services. While the rallying cry of "anytime, anywhere" access to information and electronic services has spurred a great deal of interest within the ubicomp design community, our interest is in exploring and supporting the distinctions between spaces that this approach often erases. This is ubicomp not so much anytime, anywhere but rather "right now" and "right here."

Certainly, the rise of "cell-phone-free zones" in public spaces, or cell site dampeners deployed in churches and restaurants, suggests emerging norms about the device use in difference spaces—norms that may, of course, be acknowledged and enforced socially rather than technologically. Yet the notion of seamlessness fails to acknowledge the subtle and intricate ways in which social practices negotiate, defend, reinforce, and recognize a range of boundaries between settings and spheres of activity (Zerubavel 1991). Christena Nippert-Eng (1996) points to how distinctions between home and work, as different spheres of activity with different responsibilities and forms of engagement, are maintained, managed, and navigated. Again, these are situated within different broader contexts. It is not simply that people behave differently in different spaces; being able to act differently in difference spaces, and being able to demonstrably recognize and respond to the differences between one setting and another, is part and parcel of what it means to be a competent member of society. The problem with technologies that erase these boundaries is not just that they themselves fail to recognize socially relevant distinctions but that they undermine the mechanisms by which members of society can prove, to each other, their sensitivity to these nuances.

We therefore are supportive of calls by some ubicomp researchers for "seamful" design (Chalmers and Galani 2004). Alternative approaches attempt not to erase the boundaries between settings but instead allow the technology to make visible the boundaries and seams. This is not to replace the social and cultural negotiation of boundaries but rather to enable it.

Finally, we want to point to the importance of a cultural framing of space. The experience of space, as we have observed, is coextensive with the cultural practices of everyday life, and these then provide people with a critical interpretative resource in engaging in collective action coordinated in shared spatial environments. The transformation of space through the introduction and diffusion of pervasive computing technologies must be seen in this context; not only do these technologies transform the "cultural work" being done in space, but they themselves are also sites of cultural production.

Consider two examples, again drawn from the Asian context. In the mid-1990s, the prime minister of India declared that the Internet was the future of India. This was seen by the Indian middle classes as a rallying cry and a directive, and also understood to be their personal responsibility to materialize this declaration as a reality. Many middle-class households purchased personal computers at the decade's tail end, driven in no small part by this injunction, and attempted to get online, fighting both poor telephony infrastructure and unreliable electricity. The Indian middle classes represent only a small proportion of India's more than one billion citizens, though, and access for the wider population was not a problem solved by household-based consumption. In recent years, the Indian government and various nongovernmental organizations (NGOs) have rolled out kiosk-based web services in rural villages and towns all over the country. These kiosks offer a variety of services, including caste certification, remittance payments, crop diagnosis, and access to other online government and commercial services. Experiments with novel interfaces, user models, and nontextual input are all under way, as are alternate fuel and power models. In 2008, the government, in partnership with industry, launched a new broadband initiative, the 511-Vision or Connected Indians—with the goals of getting five hundred million Indians connected and online, establishing one hundred million broadband connections, and setting up one hundred million broadband-enabled devices.

In early 2004, a subsidiary of LGE, a Korean-based consumer electronics company, launched a new mobile handset in the Middle East and southeast Asia. The handset, the Ilkone 8000, claims to be the first cell phone that supports one's spiritual practices wherever one goes. It is a 3G handset with GPS technology, a smart digital flash card, polyphonic sound, and sophisticated calendar software. This phone can find Mecca from five thousand cities around the world, and it can notify users of *salat* (prayer time) in those same locations; it contains the entire Koran in Arabic and English and can read it to the user; and it can bring the user the call to prayer from Mecca, live.

Viewed with Western eyes, what is noteworthy about this example is the relationship between religious practice and technological modernity that it embodies. In contrast to the rationalist, empiricist, and positivist narrative that underscores Western technological development, this alternative formulation embraces an alternative account of modernity in which religious practice retains a central place (Blank 2001). Religious practice is in fact an integral part of the encounter with pervasive computing technologies in a range of settings, from Buddhist blessings of cell phone handsets to SMS missives from the Pope—a service that had garnered three million users in Italy in its first two months (Bell 2006a). In the United States, 68 percent of wired Americans report using the Internet for religious purposes (Larsen 2000).

The power of infrastructures is their ability to reconfigure the relationship between the local and global. The power of pervasive computing, too, lies in this relationship, and in the ability to transform it as well. The Ilkone, like the orientation of a mosque, provides a way of making sense of the local environment in terms of its connection to Mecca and the global practice of Islam. The availability of wireless networking—whether for cellular telephony, digital communication, radio-frequency identification (RFID) product tracking, or environmental monitoring—imposes a new set of globalisms through which the local can be read, connecting one to a range of diffuse infrastructures and, through them, to a set of practices and a set of people brought instantly "into range" if not directly into view. Mizuko Ito and Daisuke Okabe's (2005a) study of mobile communication technologies in Japan highlights how technologies can be used to create new spaces for intimacy, new locales for interaction, between young married couples for whom a private physical space is economically unattainable. The technology is a means by which an intimate local space can be superimposed on, not carved out of, a global public sphere. By creating both connections and boundaries, pervasive computing technologies and the practices within which they are embedded provide an opportunity to reconfigure this local/global relationship. One interesting issue to explore, then, is the impact of ubicomp technologies on questions of scale and scalar structuration (Marston 2000; Brenner 2001); as a new set of infrastructures link individual practice to broader spatial and temporal patterns of information and activity, what new scalar configurations and "fixes" (Brenner 1998) emerge?

Infrastructure and Failure

When we link infrastructure and experience it becomes clear that infrastructures condition our encounters with the world in the course of our use of them. Infrastructure acts here as a lens through which the world is encountered but remains in the background. Still, we would be remiss in ignoring the other major case in which infrastructures structure experience of the world. The casual disattention with which infrastructures are unnoticeably put to use is most obviously manifest in our responses and attitudes to the moments in which they fail.

It is in the nature of infrastructure that it connects the mundane to the transcendent, so that infrastructural failure is often marked by sudden and discomforting juxtapositions. A major blackout of New Zealand's telecommunications infrastructure in 2005 came to be blamed, in part, on a rat chewing through a fiber-optic cable (Ward 2005); similar stories can be told involving lightning strikes on satellite base stations or damage wrought by construction (see, e.g., Neumann 1995). Similar risks attend the use of information rather than physical infrastructures—engineering enterprises that fail due to mismatches between metric and imperial measurements (Euler, Jolly, and Curtis 2001), typographical errors in software programs

(Neumann 1995), or the mundane problems of filing and finding documents (Dourish, Lamping, and Rodden 1999). On these occasions, two things are telling. First, the fragility of infrastructure—the variety of ways that problems and failures develop—throws into sharp relief the complex processes of holding it together and highlights how active that process is; stabilizing infrastructure is an ongoing effort. Second, the unexpected connection between different elements of the everyday world—rats and telephones, farming implements and the Internet, and rulers, tape measurers, and spacecraft—exposes the complexity of maintaining infrastructure, both in a pragmatic sense of protecting and tending to physical structures and an ontological one in the substitution, as a matter of practical engagement, of the abstraction that infrastructure offers from the complexities of its implementation. Both take work.

At the same time, failure reveals itself in the small. For every spectacular failure of infrastructure there are hundreds or thousands of small everyday niggles and petty failures—doors that jam, mobile phones that need to be power cycled, keys that require a special touch, antennas that need to be aligned just so, screws that refuse to turn, wireless networks with inadequate coverage, operating systems that lose track of memory, plumbing that backs up, or cables that need to be jiggled. Many of the indignities of daily life revolve around the continual mutual alignment of individual action and infrastructures that don't quite support it, either because they do not operate entirely as advertised or because they have been pressed into service to fulfill needs that they were never intended to support. Elsewhere, we have suggested that visions of technological futures enshrined in science fiction can frequently be distinguished by their orientation toward infrastructural failure as dramatic and spectacular, or as mundane and unremarkable (Dourish and Bell forthcoming).

It is perhaps this picture of niggling failures that contributes to the denial of ubicomp as a contemporary reality—a topic that we explore in the opening chapter. Ubicomp is still a thing of the future, this argument goes, because our current approximation is still troubled by failure and discontinuity. If, though, we take failure and discontinuity as inevitable consequences of any technology, and as aspects of everyday life that can't be eliminated, then the picture changes somewhat.

First, as we outlined previously, it suggests that we should think about ubicomp not as a project to be completed but instead as something that already has a present-day form and can be studied as a contemporary phenomenon. This is not to say that it will not change, of course, but to assert that contemporary practice is worth examining even if it does not look the way we imagined. This is a point we have made before.

Second, and consequently, when undertaking these kinds of investigations, it focuses us not on the ways that ubicomp infrastructures will seamlessly and unproblematically support human activity but rather on the processes by which human activities are aligned with infrastructures that themselves are fragile and brittle. In other words, it turns our attention to ubicomp as a process and practice as opposed to an object.

Third, it suggests that the failures, niggles, problems, and breakdowns of contemporary ubicomp are not things to be written out of the picture, on the assumption that they are temporary problems to be overcome in the next generation of ubicomp (the "real" ubicomp of which our current implementation is merely an approximation). Instead, while the details of the technology change, the notion of everyday practice as needing to work around infrastructure as much as through it seems likely to remain salient.

Ubiquitous Computing Is Messy

Our goal in this chapter has been to examine two aspects of infrastructure and practice relevant for emerging ubicomp technologies and environments; we label these the infrastructure of experience and the experience of infrastructure.

In talking about the infrastructure of experience, we have concentrated on how our encounters with everyday environments depend on both the practices in which we might be able to engage and the structures that are inscribed into those environments by those practices. The experience of space, we have argued, is coextensive with the cultural practices of daily life; those practices, in turn, provide the framework through which space is experienced and rendered locally and collectively meaningful.

In talking about the experience of infrastructure, we have concerned ourselves with the ways in which infrastructures offer themselves up to people for manipulation and interaction. Infrastructures, normally taken for granted and an unspoken part of the background, must nonetheless be managed, negotiated, navigated, and made to work as a part of the environments and practices that they support. Infrastructures can recede into the background—that is, exist only in the context of well-understood practices and only through continual efforts of management and maintenance.

Of course, the relationship between these two elements is recursive; infrastructures give meaning to experience and experience gives meaning to infrastructures. The reason to take this infrastructural approach is precisely because infrastructures come with "points of view"; as Star (1999, 379) notes, "Study a city and neglect its sewers and power supplies (as many have) and you miss essential aspects of distributional justice and planning power." This provides us with a useful entrée into questions of design practice. From this perspective, we draw a number of conclusions with implications for the development and analysis of pervasive computing technologies and environments.

The first and most fundamental one is that space is organized not just physically but also culturally; cultural understandings provide a frame for encountering space as meaningful and coherent and for relating it to human activities. Technological infrastructures are inherently given social and cultural interpretations and meanings; they render the spaces that they occupy ones that can be distinguished and categorized and then understood through the same processes of collective categorization and classification that operate in other domains of social activity. Technological infrastructures and services, then, need to be understood as operating in this context.

The second is that architecture is all about boundaries and transitions and their intersection with human and social practice. We need to think architecturally about the mobile and wireless technologies that we develop and deploy, the human side of infrastructures. Everyday spaces are not simply spaces for working or meeting but also for waiting, reading, loitering, watching, loving, remembering, and more (McCullough 2004). The rhetoric of seamlessness is often opposed to the inherently fragmented nature of social and cultural encounters with spaces; we need to be able to understand how ubicomp might support rather than erase these distinctions.

The third is that new technologies inherently cause people to reencounter spaces. This is not a question of mediation but one of simultaneous layering. One fascinating aspect of the move from the systems we built on the wired Internet to those that we experience through wireless and mobile networks is that we are creating not a virtual but instead a thoroughly physical infrastructure, and we need to think about it as one that is interwoven with the existing physical structure of space (Dourish 2001). The rhetoric of ubicomp is one that traditionally ignores the ways in which computing experience must be implemented on top of, and experienced in and through, an existing landscape, whether that means Wi-Fi hot spots on a university campus (Barkhuus and Dourish 2004) or non-Western domestic topologies (Bell 2006b).

Finally, there is already a complex interaction between space, infrastructure, culture, and experience. The spaces into which new technologies are deployed are neither stable, uniform, nor given. Technology can destabilize and transform these interactions but will only ever be one part of the mix. Accordingly, the goal of pervasive computing must be to design not simply for settings but also for the processes by which practice and meaning evolve. Ubiquitous computing was, from the outset, a proposal not for how technology should *be* but instead how it should be *experienced*.

6 Mobility and Urbanism

As we discussed in earlier chapters, one of the predominant characteristics of contemporary ubicomp is its fascination with mobility and mobile devices. That mobility is part of the ubicomp research agenda is hardly a surprise; it was prefigured in Weiser's first ubicomp scenario and has been a consistent part of most research projects since then. The nature of mobility, and the ways in which it is expressed in devices, technologies, and practices, however, has undergone subtle and not so subtle transformations. In that first scenario, Sal's car features onboard computation, and responds in real time with location and commerce information. There is no accounting for mobile phones, or other forms of handheld or portable computation. In 1991, computing was still something linked primarily to particular places (i.e., academic labs and computing centers), and the dominant paradigm for information services was the desktop computer connected to fixed infrastructure. Weiser wanted computing to disappear into the walls and furniture; he did not imagine it untethered in our pockets and purses. But today, many researchers in ubicomp feel that the mobile phone should be the primary platform for deploying new information services (Satyanarayanan 2005; Patel et al. 2006).

In many ways this reflects the significant shifts in the technological landscape over the last two decades: there have been widespread adoptions and uses of networked technologies and, most especially, deployments of wireless and mobile systems via wireless Ethernet and cellular telephony systems (or there should have been). These are not simply technological transformations but also transformations in social and cultural practice; our expectations about what computers are, what they might do for us, and the role of digital objects in everyday life have evolved considerably in the past nearly twenty years. The same period has also seen significant shifts in the workplaces in which ubicomp was imagined; if the organizational rhetoric of the 1980s centered on knowledge management, core competences, and communities of practice, that of today is more concerned with virtual teams, global connectivity, and service delivery. Mobility—of people and activities, on a large and small scale—plays a different role.

In other words, we need to look beyond a technical and instrumental account of mobility. When we think of mobility in strictly technical terms, we think of topics such as bandwidth, resource management, location, and wireless networks. But when we think of mobility in social or cultural terms, a different set of topics come into view: pilgrimage and religious practice, globalization and economic disparities, migration and cultural identity, and daily commutes and the suburbanization of cities. How these various elements are configured and reconfigured is subject to constant negotiation, including even the ongoing one of individual identity. Different places demand different forms of behavior, yet mobile technology suggests the possibility of a constant, fixed expectation about who one is and one's level of accessibility and engagement.

In this chapter, we examine the cultural contexts of mobility and read these back against the dominant tropes and themes of mobility within ubicomp. This argument extends and develops recent attempts to explore the social and cultural aspects of mobility in HCI and ubicomp (Bidwell and Browning 2006; Brown and Perry 2002; Dourish 2006b; Ito, Okabe, and Matsuda 2005; Turner and Davenport 2005), conjoined with a burgeoning interest in collaboration and interaction beyond traditional working settings, including leisure settings (e.g., Agostini et al. 2002; Brown and Chalmers 2003; Ducheneaut and Moore 2004; Reeves et al. 2005) and museums, or gallery spaces and universities (e.g., Heath et al. 2002; Hindmarsh et al. 2002; Burrell and Gay 2002; Hornecker and Buur 2006). As we have done throughout this book, we draw on a range of accounts that lie outside the domain of information technology to gain perspective on technological design. Exploration in areas such as cultural geography and the anthropology of space suggest alternative accounts of mobility that should prove fruitful for ubicomp. We begin here by discussing the relationship between mobility, spatiality, and technology as it has developed in ubicomp research before presenting a series of contexts—mythical, moral, imagined, and historical—for everyday mobility, and reflecting on what these might tell us about the prospects for mobile interactive technologies. We use recent projects to illustrate new avenues for design exploration and analysis.

"You Couldn't Tell Pitt Street from Palm Sunday": Defining Our Terms

Sociologist John Urry (2000, 2007) has suggested that mobility, rather than society, may be the primary animating metaphor for sociology in the twenty-first century. He argues that in contrast to a traditional sociology focused around the emergence, maintenance, and character of social stability, contemporary sociology must increasingly contend with mobilities of many sorts—movements of people (migrations, diasporas, tourism, jet-setting, business travel, suburbanization, commutes, and more) but also of goods, capital, information, and media—and their dynamics. His assertion cuts both ways: not only does it indicate that mobility must be an important concern for social analysis, but it also highlights the importance of social analysis for any account of mobility. In a talk at the University of California, Irvine in 2007, Yale anthropologist James C. Scott remarked that the major topic of his life's work has been why people who move around find themselves enemies of those who don't. Scott's work over the last twenty-plus years with landless peasants in Asia (1985, 1998, 2009), migrant workers, and hill people has meant repeated encounters with populations that were highly mobile and thus difficult to regulate, monitor, and control; that all ultimately posed significant problems for their various nation-states. Indeed, contemporary celebrations of the road warrior are radically at odds with a longer-term picture of the troublesome nomad, whether that manifests itself as the questionable moral status of the medieval troubadour, the nineteenth-century U.S. "tramp scare," or contemporary Western debates about immigration and asylum seekers (Cresswell 2006). Mobility is not simply movement from point A to B. Transnational migrations, economic globalization, and religious pilgrimages are obviously forms of mobility that need to be understood socially, but so too is the daily commute, the venture downtown for an evening's entertainment, or the vacation.

What this implies is that when we approach the topic of mobile technologies, we find ourselves at the nexus of two powerful intellectual and cultural currents. Both mobility and technology are deeply embedded in particular ways of thinking and in imagining the world and ourselves. In this chapter, our goal is to examine the relationship between these two concepts and especially how ubicomp might account for the many different, simultaneous cultural embeddings of mobility.

It is precisely in relation to these issues of connection and disconnection, or uniformity and difference, that mobility and mobile technologies play such a key role. Patterns of connection arise around forms of movement and mobility; our sense of spatial organization emerges from the patterns of movement of everyday life, as made visible in Kevin Lynch's study (1960) of people's "egological" maps of their cities. At the same time, technologies of all sorts—from passenger jets and handheld GPS to maps—give us ways of seeing places as connected or unified. Technological infrastructures create seams—the boundaries between zones of operation and non-operation, regions of stability and difference (Chalmers and Galani 2004; Dourish and Bell 2007). We start to navigate and understand spaces in terms of the manner in which they reveal themselves to us as we move through them and in terms of the portability of practice as we move—whether we're within range of a cell phone tower, a Wi-Fi zone, or a power outlet. New technologies, new technological practices, and new forms of mobility are associated with new ranges of questions about sameness and difference that give everyday space structure and meaning. If there is cellular coverage, is it 3G, EV-DO, or EDGE? If there's Wi-Fi coverage, is it with the right provider? If there's no Wi-Fi, where in town is access available? Mobile technology is not, then, simply operating within a specific environment; it is implicated in the production of spatiality and spatial experience.

Mobility and Urbanism: Some Examples

Early work on mobility in ubicomp and related fields focused on the mobility of office workers, especially within workplace settings (Bellotti and Bly 1996). As publicly available infrastructures were deployed that allowed online activities to move beyond the walls of offices and labs, the city became a major concern for a developing area of "urban informatics" (Foth 2009).

Technology and the urban experience have always been deeply entwined. The city embodies modernist ideals of rational planning and the products of industry harnessed for the benefit not just of the individual but for civil society as well—a model of technology's application to the fabric of everyday life. The city is not just a concentration of humanity; it is a nexus of technological infrastructure, on which it depends for its continued existence. It is perhaps unsurprising, then, that the rationalist city would feature as a site for ubicomp. Wayfinding, people finding, time management, and resource consumption as parts of city living emerge as major topics of technological consideration (e.g., Bedwell et al. 2009; Bellotti et al. 2008; Stewart et al. 2008). Information technologies, particularly those of mobile networking and positioning, become a new lens through which the spatialities of urban space can be viewed.

Can You See Me Now, an urban street game that overlays two spaces—one online and one real—exemplifies these overlapping spatialities (Flintham et al. 2003). Originally deployed in Sheffield, England, in 2001, it allows online players connected to the system through a website to move their virtual representations through a map, while human players armed with GPS location devices and wireless Internet connections hunt them on the streets of the real city. When the human and virtual players reach the "same" spot, the virtual player is caught. While this layering of spaces itself is intriguing, what is interesting for our purposes is the forms of tactical play that emerge in how virtual players exploit their knowledge of the city's topography (busy streets to be crossed or steep hills to be climbed), even as the human players similarly begin to exploit GPS black spots as well as areas of high and low network connectivity and to incorporate these into their gameplay (hiding in GPS black spots in order to ambush unsuspecting virtual players, for example). In other words, the presence or absence of network services become a new way to reencounter the city streets. This is not simply thought of as an overlay of the virtual and physical; GPS satellite line of sight and Wi-Fi network signal strength are thoroughly physical phenomena. What they do provide, though, is a new way to think about the space and what one can do there—a new mobility of access, presence, and interaction.

On a broader scale, a number of researchers have investigated the role of maps and spatial practice in tourism (e.g., Brown and Chalmers 2003; Norrie and Signer 2005). Tourism brings these issues to the fore because of the limited spatial resources available to tourists and their inherent unfamiliarity with the spaces through which they move. The very existence of tourists and tourism as a category—and a target demographic for technology developers as well as "urban entrepreneurs"—is dependent on specific conceptions of space, culture, and nature, hence the flows of people through and between "cultural" and "natural" spaces (Cronon 1996; Smith 1989). Tourism defines a relationship between an individual and other tourists in whose footsteps one moves and with whom one travels, either formally or informally. When we speak of "tourist trails," we are speaking almost literally, and both paper and electronic maps for tourists do not just document but also produce spatial forms and spatial experience.

The Urban within Ubcomp

Cities are enmeshed in these flows and reflect them in their own structures. As we have suggested in the prior chapter, the infrastructure of the city enables, hinders, and directs these flows, resulting in an experience of the city that is both heterogeneous and dynamic. The daily rhythms of everyday life (transportation, the movements of people and goods), the longer-scale evolution of cities (outward expansion, the flow of people between urban centers and suburban and exurban communities), and the broader historical patterns that link places together are all of interest and importance.

With the seeming ubiquity of information and communication technologies has come an emphasis on what may seem to be a collapse of time and space. Castells (2000) discusses globally distributed processes of production, in which capital, labor, management, and markets may be half a world away from one another yet linked and coordinated. Harvey (1989) uses the term "time-space compression" to describe the processes at work, which he argues are inherent to contemporary capitalism. In his formulation, an intensification of temporal pace—twenty-four-hour production models, speedier processes of obsolescence, and a quickening of the business cycle—is combined with a lessening of the impact or obstacle of distance, both through the use of digital technologies to bridge distance and the global distribution of commercial and industrial operation. Information technology clearly plays a key role here, although it is crucial not to assume a causal one; as Harvey points out, capitalism is inherently expansionary, and we could as easily see this need to expand as a driver of technological innovation rather than presuming that technology is the spur.

We find these considerations of global flows and patterns especially significant when we start to think about technology's role in urban experiences (e.g., Mainwaring, Anderson, and Chang 2005). Several issues present themselves.

The first, which will occupy us further in a moment, is the role that technology performs in staging encounters between people and urban space. Transportation systems are the most visible of these, as noted in the previous chapter and as illustrated by Janet Vertesi's exploration (2008) of the London Underground map as a mental model for the organization of London. Nevertheless, with the increasing interest in municipal networks, we start to encounter the spatiality of the city through the range of services that might be available there, especially when such services are deployed selectively. An investigation of the spatial correlation between Wi-Fi access points and median household income might be instructive.

The second consideration is the very way in which we think about personal mobility and urban movement in the context of technology design. As has been noted elsewhere, the cities that are the sites of urban computing research are typically quite similar. First-world "world cities" with significant infrastructures and capital investments, like San Francisco, New York, London, and Tokyo, feature prominently; strangely absent are such metropolitan centers as Kuala Lumpur, São Paolo, Seoul, Singapore, Detroit, or Calcutta (Dourish and Bell 2007). We would note further that the contexts of mobility have been similarly constrained. The urban resident is frequently pictured as young, well-heeled, techno savvy, and above all engaged in discretionary movement through and consumption of urban space. The overwhelming sense that urban computing technologies convey is one of options and opportunities. Urban computing technologies help people answer questions like: Where shall I go today? What's the latest "happening" restaurant? Where might I find people whom I might like? They encourage an appropriation of space (Chang and Goodman 2004) in ways that certainly reflect Michel de Certeau's concern (1984) with spatial tactics, but they often fail to acknowledge his considerations of the systems of power and control within which those tactics emerge (and against which they should be read). By way of contrast, let us think of other residents of urban space whose orientation toward mobility might be quite different—the homeless, for whom movement is a way of avoiding problematic encounters with authority (Bittner 1967; Spradley 1988); parolees, whose movement is constrained and monitored (Troshynski, Lee, and Dourish 2008); taxi drivers, for whom mobility is a form of labor (Davis 1959); or low-wage employees who spends upward of four hours a day on public transit to reach employment to support themselves and their families (Brewer and Nguyen 2006; Brewer, Mainwaring, and Dourish 2008).

These different circuits of mobility intersect and overlap in urban spaces and provide alternative opportunities for thinking about the spatialities of technology. Eric Paulos and Tom Jenkins's engaging "urban probe" (2005) encourages people to reflect on the patterns of occupation and action of urban space seen through the lens of the trash can; yet it fails to incorporate the idea that for a small but significant number of urban dwellers, the public trash can is a source of food.

Alternate Approaches to Mobility

As we suggest in earlier chapters, particular ways of knowing the world around us are mediated cultural lenses, such as in our discussion of Warlpiri mapping. In order to gain a broader view, we explore some alternative cultural logics of movement and mobility. We want to step away from the specificities of mobile technologies for a moment to look at how mobility is culturally constructed and grounded in specific locales and geographies. Using a range of ethnographic and historic examples, we tease out several critical areas around which one can talk about mobility beyond the current ubicomp frame: social practice, morality, the imaginary, and the historical.

Mobility as Social Practice

Over the past decade or so, mobile technologies have been more pervasive outside the United States than in it. The adoption of mobile telephony on a global scale has been mind-boggling. Almost ten years have already passed since the number of landlines in Spain and Italy were eclipsed by mobile ones; nearly two-thirds of the populations of China and India now own mobile phones, and many more people in these countries count themselves as regular users. These rapid rates of adoption reflect an interesting constellation of factors: existing patterns of high (albeit localized) social mobility outside the home, strong government regulation mandating pan-national standards (i.e., the European Union), poor existing terrestrial telephony (i.e., in India and China) and good national cell coverage (i.e., in the United Kingdom), and competitive and strategic calling plans and pricing policies.

In the late 1990s, Bell and her colleagues from Intel's Peoples and Practice Research Group conducted multi-sited ethnographic fieldwork in Italy, France, Spain, Germany, and the United Kingdom, exploring notions of social mobility in middle-class lives across a range of urban, suburban, and semirural settings (Bell 2001; Sherry and Salvador 2001). The project was framed to make sense of the different trajectories of mobile device uptake. One of its key findings was a strikingly simple one: mobility was inflected through a profoundly cultural lens—what it means to be mobile has distinct cultural meaning. Not only were there strong differences between the European cultures—what constituted mobility in Britain, for instance, was quite different than in Italy—but there were also strong differences between the broader European traditions around mobility and those of the United States. Much of these differences turned less on mobile practices themselves than on the dialectic relationship between mobility and domesticity, where mobility was frequently understood as being away from home. To put it another way, the construction of home and public, and the relationships between them, also has impact on the adoption and use of mobile devices.

In the United States, a great deal of social activity and intimacy goes on within the home. Within certain demographics, the locus of social activity is entirely domesticated. In other cultural traditions and sites, however, there is a great deal more slippage between the notion of social activity and the appropriate site where such an activity might take place. There is a range of sites through which people transit and linger and a variety of ways through which such sites are connected and disconnected from the home. This sort of social mobility plays an enormous role in the uptake of mobile devices and the uses to which such devices are put.

In Bell's research, we heard a great deal about what it took to be mobile in the early twenty-first century. It necessitated infrastructure in various forms: transportation (train lines, highways, cars, bikes, footpaths, etc.), utilities (phone lines, electricity, etc.), high technology (mobile phones, laptops, phone cards, credit cards, etc.) and low (wheeled bag, phone cord, etc.), accessible consumer installations (Kinko's, Gap store in every town, etc.), resting and refueling points (gas stations, fastfood restaurants, hotels, ATMS, cafés, beer gardens, mobile phone stores, etc.), and personal/personnel (travel agents, administrative assistants, spouses, family members, etc.).

At the risk of overgeneralizing, the research indicated that in the United Kingdom, for example, a great deal of time each day is spent away from home. Indeed, it is fair to say that the British home exists within a set of competing hubs of social activity, including the "local" pubs, gardens, community centers, and sporting facilities. The British home is also located within a constellation of sites of work, consumption, and transportation. People move between these local hubs, frequently without first going home, and it is easy to see why mobile devices (initially phones) might map well onto this pattern of mobility. Here the mobile phone and other mobile devices become tools for arranging social activity and touching base as much as they are for productivity and business. Furthermore, it is possible to see how Internet access points (in the form of Internet cafés and phone-booth web kiosks) would and have flourished in ways that have not been so successful in the urban United States.

This concept of social mobility also implies a flexible notion of geography, where distances can be collapsed or negotiated. Some forms of social mobility are local. For instance, ethnographic research in Spain identified the ways in which the barrio and plaza operate as an extension of domestic spaces, through which children circulate under the watchful gaze of a whole community of parents (Bell 2001; Low 2000). Others imply wider networks and spaces. The relationship between Paris and all regional centers within Brittany was one of center to peripheries, where Paris was an aspirational end point of a certain kind of social mobility—both economic and cultural. Social mobility recalls the particulars of history and geopolitics too. The Berlin of 1999 still had the distinct markings of the previous lines of demarcation, and residents vividly recalled the time when governmental authorities actively discouraged mobility beyond well-defined and heavily policed boundaries; certain sites were off-limits, and social activities and social circles were prescribed.

Mobility and Morality

Like the Warlpiri case we discussed in early chapters, Native American practices connect physical spaces to cultural values, but in different ways. In what we present here, we draw on studies of the Western Apache by Keith Basso (1988, 1996).

Basso's account of the role of the landscape is strongly connected to the stories told about it. Stories play a prominent role in Native American cultural practices, conveying history as well as moral and practical lessons. They are, in fact, the primary ways in which these lessons are transferred, and everyone learns a stock of stories and their implications. Wisdom, for the Western Apache, consists partly in being able to draw appropriately on a collection of stories in order to illuminate current situations. Being sensitive to these resonances, and being able to draw appropriately on the stock of stories to understand current events, allows the wise to avoid potential mishaps.

Further, stories are set in places; placeless events are nonsensical. Indeed, the names of those places may be used to refer both to the stories and the moral lessons that they embody. The places where stories happened are always reported and allow people to imagine themselves there, since these places are not imaginary or distant ones but instead are generally known to the people who tell and hear them. In line with their view of conversation as a cooperative effort, a good storyteller among the Western Apache does not talk a great deal; being a good storyteller rather involves "giving people pictures," helping people to visualize themselves in places, where the story may play out in front of them. These places (and particular spatial orientations and vistas) are so central to the telling of stories that their names may become proxies for the stories themselves. It is not just that stories are about places, then, but that they are about *being in* places.

For example, Basso provides a number of examples in which people comment on a current event by listing a series of place-names ("It happened at Line Of White Rocks Extends Up and Out, at this very place! It happened at Whiteness Spreads Out Descending To Water, at this very place!") that are not intended to situate the events spatially but rather to draw others' attention to the analogies between the current situation and the stories of the events that happened at those places. The names of familiar places stand for the stories and lessons that the stories contain. The landscape is central to the telling of stories and central, therefore, to the work that stories do.

Certainly, the stories reflect practical lessons (where to get water, where to find shade, where to plant crops, etc.) that make an unforgiving landscape a little easier to navigate and inhabit. But the work that the stories do is also moral. The stories capture the lessons of experience (either historical or mythical), and through their repeated retellings, reinforce conventions of appropriate behavior and social norms. Again, the places where the stories take place are generally local ones, well known to those who use the stories. They are part of daily experience. Consequently, a strong connection is formed between the moral foundations of the social order (as expressed in the stories) and the everyday experience of space. In one telling example that Basso relates, a girl talks of being "stalked by a place"; she refers to the shame that she feels when she passes a particular place, which was the setting of a story told once by an elder woman to upbraid her for her inappropriate behavior. The elder woman had told the story in order to note the girl's deviation from communal standards of behavior (in this case, how she had worn her hair at a recent ritual), but what is especially interesting for our purposes is how the landscape served to reinforce this lesson, because the story and place are so tightly bound together. In these settings, then, the physical world becomes a moral landscape, reflecting collective standards of behavior, embodying lessons, and forming the basis of wisdom.

Not only are spaces morally grounded, but navigation through, among, and with respect to them has a moral character. Here we literally see the notion of a "moral compass"—the ways in which movement through and situated within particular spatial environments links one into a set of collective values.

Mobility Imagined

Our third example concerns our understanding of the structure of space for and through navigation. We have already seen, in the instance from Munn's work in chapters 4 and 5, how cultural experiences of space may include the ways in which that space can be traversed. We argue that this is in fact a bidirectional link; not only do we draw on cultural and historical meanings in order to find space traversable, but the manner in which we navigate space gives rise to the structure we find in it.

One compelling case is presented in Daniel Lord Smail's study (1999) of the emergence of street addressing in late medieval Marseille. Through a careful analysis of archival records and the forms of location specification they exhibit, Smail outlines four ways of describing specific locations. One is by reference to landmarks—churches, statues, prominent citizens, civic buildings, topological features, or other remarkable characteristics of the city's geography. Paths or proximity to these landmarks are often sufficient to note locations. A second form of reference is with respect to particular districts, such as church parishes, which provide a patchwork of regions within the city. One interesting class of districts is "islands"—essentially entire city blocks (that is, sets of buildings bounded by streets on all sides) conceived of as coherent units, and frequently labeled with reference to their inhabitants and the activities found there (e.g., the island of the shoemakers). A third is what Smail terms "vicinities." Like districts, these name areas or regions of the city, but unlike districts they have no formal or physical boundaries but are informal designations that are familiar to residents. Often, these refer, again, to the artisan practices to be found in that area. One interesting issue, though, is that the craft associations with a particular region of the city are sometimes historical rather than contemporary; that is, the "quarter of the goldsmiths" might name an area where goldsmithing had traditionally been carried on, even if it were no longer a dominant craft there. Understanding these references requires a knowledge of not just the geographic but also the historical specificities of the place.

Streets as we understand them are the fourth of Smail's categories, but they are by no means the dominant form of addressing in the documentary record. (It is worth noting that Smail's records are notarized contracts; so he is talking about formal documents, not vernacular practice.) Over time, however, street references overtake the other forms to become the primary means to identify locations. Smail connects this to a number of associated transformations, including changes in the legal system and political order. One fascinating association is with the notaries recording the contracts whose mental cartography became more street-oriented as they themselves became increasingly mobile. In a context where few people moved through the city extensively, those who did came to think of the space as one to be navigated; hence the primary mechanisms of navigation, the streets, became their ways to fix location. The move from islands to streets is essentially a figure/ground reversal that we can associate with the conception of the city as something to be occupied or moved through, suggesting that streets become prominent features first for navigation and only secondarily as a means to locate specific buildings or people. It is mobility that undergirds a notion of the organization of space; space is understood differently as patterns of movement change.

What we see in these cases is that patterns of movement and mobility in urban environments become the basis of finding that space meaningful. Transportation systems may provide orienting axes as the "user interface" to the city, as Vertesi (2008) puts it. More broadly, mobility lends the city a social geography.

Mobility in History

Finally, we look specifically at the ways in which history plays a role in constitutions of mobility. Here again it is useful to invoke Massey's concept (1993, 61) of power geometries, as mentioned in chapter 2:

> For different social groups and different individuals are placed in very distinct ways in relation to these flows and interconnections. The point concerns not merely the issue of who moves and who doesn't, although that is an important element of it; it is also about power in relation to the flows and the movement. Different social groups have distinct relationships to this anyway-differentiated mobility: some are more in charge of it than others; some initiate flows and movement, others don't; some are more on the receiving end of it than others; some are effectively imprisoned by it.

Power geometries thus describe the ways that spatial arrangements (e.g., the locations of homes and their proximity to both amenities as well as sources of noise and pollution) along with patterns of access and mobility (e.g., in the competition for resources between different forms of public and private transportation) reflect arrangements of power and control—arrangements that are historically constituted on both a small scale (such as the patterns of stratification within urban space) and a large one (such as the echoes of empire in the multicultural landscape of the cities of both former imperial powers and former colonies). More broadly, navigating and understanding space involves an orientation toward the social structures encoded within that space.

Historical aspects of navigability can be found written into other addressing systems, especially those in which regions or houses are numbered in order of construction and where traditional patterns of occupancy (e.g., associated with patterns of immigration, employment, ethnicity, or lifestyle) give rise to different forms of spatial segmentation (Lyons 2003).

William Kelleher's study (2003) of memory and identity in Northern Ireland highlights the experience of spatial arrangements reflecting sectarian and political divides. In his inquiry, the division between Catholic and

Protestant is a central feature of everyday life. As local residents took Kelleher on tours around their town and its surrounding area, they articulated the space in terms of sectarian associations that mix religious, political, and historical elements. Particular regions of the town are read as being associated with one group or another, in terms of current occupancy and historical patterns of migration. People orient toward the space they occupy and navigate in terms of the social organization of daily life, in which these distinctions play a central role. To be in particular places at particular times is to mark oneself as being a particular sort of person—or potentially to place oneself at grave risk. Current and historical patterns of settlement create navigational "fault lines"—invisible barriers that are rarely crossed. The understanding of "our" space and "their" space is a daily aspect of experience. Certainly this goes well beyond encounters with space; on a more detailed level, even patterns in housing stock have a sectarian reading, as do body images, clothing choices, and forms of talk.

Here, then, the encounter with space is also an encounter with social structure, its antecedents, and causes. Other work has pointed to the ways in which such spatial distinctions have their correlates in social networks—an interesting observation in light of the recent technological interest in mapping and articulating social ties (Grannis 1998).

Rethinking Mobility (and Urbanism) in Ubicomp

Our choice of examples here is motivated by what George Marcus and Michael Fischer (1986), channeling Viktor Shklovsky (1925/1990), call "defamiliarization"—an attempt to explore alternative cultural practices in order to gain perspective on more familiar, everyday considerations. So by turning to questions of spatiality in a range of different cultural settings, our goal is to highlight common patterns across those settings. We might equally look to Charles Goodwin and Marjorie Harness Goodwin's study (1996) of the practical spatial organization of aircraft movements, Theodore Bestor's analysis (2004) of the transnational economy of culturally recognizable foods such as sushi, or Monika Büscher and her colleagues' account (2001) of spatial reasoning in the work of landscape architects to make our case.

Across all these instances, what we find is that the encounter with space is framed by cultural logics, or a series of collective understandings through which space, spaces, and their representations take on particular kinds of meaning. These logics are themselves social products; they arise out of our actions and interactions as we move around in and make sense of the world

(LeFebvre 1991). The cultural logics shape, and are shaped by, patterns of movement and action in space. Examples might include the conventions of daily domestic life (e.g., distinctions between the public and private parts of a house), the structural properties of cities and commutes (e.g., the emergence of urban/suburban regions, and the forms of activities and buildings we associate with them), or global patterns of migration (e.g., the waves of outward and inward movement associated with the colonial and postcolonial periods). The social character of spatiality and mobility arise out the confluence of these sorts of patterns.

What is especially of interest here is the ways in which information technologies provide sites and occasions for the development of new forms of environmental knowing. How does the presence of technological infrastructures such as GSM or Wi-Fi shape or respond to patterns of movement and activity in space? Certainly, those infrastructures are deployed in response to expected patterns of habitation and migration (cell phone towers are denser in urban centers and along established routes), but at the same time they give rise to new kinds of spatial understandings (e.g., seeking out coffee shops to find a Wi-Fi signal to work for an hour or two).

Technologies of all sorts—maps terrestrial, maritime, and cadastral; compasses; sextants and theodolites; steam engines; tide charts; and square rigging—have always played a key role in how we understand the spaces through which we move (Law 1987). While our concern in the examples that we have examined in this chapter has been primarily with mobility (rather than technology), it is noteworthy that the operation of the cultural logics we have explored is conditioned by the technologies through which the landscape may be encountered and navigated, including technologies of mobility as well as representation (i.e., maps and charts). Similarly, information technologies are deeply implicated in the operation and emergence of these logics, and in the forms of collective encounters with space. It is through these connections that we want to return to technological topics and focus on the lessons that we can draw from the accounts presented above.

To do so, we need to bring in a third element: not just mobility and technology but also practice. The lens of practice—how people act in space, and how those actions render spaces meaningful—provides a critical link. One interpretation of this link between cultural accounts of mobility and technology is that our interest must be directed toward how information technologies create new virtual spaces that transcend and overlay the "real" spaces of the everyday world. In fact, we would suggest that a number of attempts to create electronic spaces for collaboration and communication, such as technologies for "virtual copresence" or telepresence, have often been founded on just this sort of principle. Yet we would argue for a quite different interpretation of the relationship between place and space in technologically mediated practice. The technologically mediated world does not stand apart from the physical one within which it is embedded; rather, it provides a new set of ways for that physical world to be understood and appropriated. Technological mediation supports and conditions the emergence of new cultural practices not by creating a distinct sphere of practice but instead by opening up new forms of practice within the daily world, reflecting and conditioning the emergence of new forms of environmental knowing. Our concern is with the role of technology in spatializing practices.

Ito and Okabe's discussion (2005a, 2005b) of aspects of Japanese use of mobile telephony and messaging, noted in the last chapter, offers a series of vivid examples. Two are relevant here. First, they note the critical role of mobile-messaging technologies in face-to-face encounters in the city. Like Richard Ling and Birgitte Yttri (2002), Ito and Okabe point to the ways in which mobile-messaging technologies support "micro-coordination," providing a "last 100 yards" solution for rendezvous as well as allowing for a fine-grained coordination of actions in space when people are together. They also show, however, that mobile messaging, beyond micro-coordination, provides for different forms of presence as a part of a rendezvous. In a large and complex city like Tokyo, travel can be challenging, especially at busy times, but among the teens they studied one is not "late" to a meeting if one participates virtually. Mobile messaging is a proxy form of participation when one is not yet physically at a meeting spot; "presence in the virtual communication space is considered an acceptable form of initial 'showing up' for an appointed gathering time" (Ito and Okabe 2005b, 268).

In a second example, Ito and Okabe (2005a) discuss the use of phones to allow private and intimate communication among those who are otherwise unable to find the privacy or autonomy to maintain such relationships. Examples include young people whose mobility in urban spaces might be limited, and whose autonomy may be strictly curtailed by parents, teachers, and others; college-age adults who live at home with their parents, because of the high cost of housing, before or even after beginning to work; or young couples who find that the same housing market forces them to live apart until they have accumulated money for a larger place together. For people in these situations, mobile messaging supplies an opportunity for private communication and intimate extended copresence throughout the day.

It is perhaps tempting to see this as suggesting that new electronic spaces are being created that transcend the spatial arrangements and constraints of mundane reality. We resist such an interpretation, though. The "technosocial situations" that Ito and Okabe detail are certainly forms of social and cultural practice that rely on information technology for the forms that they currently manifest. Still, they are firmly situated within, motivated by, and shaped in response to everyday life. Mobile-messaging technologies in the examples cited by Ito and Okabe do not create new spaces but rather allow people to encounter and appropriate existing spaces in different ways (see also Ellwood-Clayton 2003, 2005; Prøitz 2005). These new mobile practices, then, transform existing spaces as sites of everyday action. Far from seeing technology as creating a space apart, we view it as being fundamentally a part of how one encounters urban space and how it is shaped through technologically mediated mobility.

In our previous examples, we tried to emphasize forms of cultural experience of space, and in so doing to motivate an approach to place and space that puts place first and understands the social as collective and embodied cultural experience. The examples provided by Ito and Okabe point the ways toward a reconfiguration of the conceptual relationships between place, space, technology, and practice as they frequently figure in discussions in ubicomp and computer-supported cooperative work. Clearly, technology—of all sorts, including information technology, construction technology, transportation technology, and more—plays a critical mediating role in our experience of space. But it does so within a cultural context that gives these technologies meaning as parts of daily life. Information technology is instrumental in producing social and spatial arrangements.

From Mobility to Fluidity

Our efforts here have been directed at infusing a technological account of mobility with a sense of the social and cultural dimensions within which those technological achievements come into being. As a technological requirement, mobility emerges as a set of constraints on forms of connectivity, size and power requirements and an attentiveness, first, to the problems of partial or intermittent connectivity and limitations of access and, second, to the opportunities associated with a model in which technology moves to the sites of human interest rather than the other way around. From a social perspective, though, the meanings, relevances, and ramifications of mobility reach well beyond those instrumental concerns. Mobility is not simply about getting from point A to B; it is a form of social practice, and the intersections of those practices with the infrastructures that support them are not straightforward.

When we take this perspective—the intersection of infrastructures that are at once technical, social, and cultural—we are led to think about space quite differently, as always comprising a historically and materially constituted hybrid of physical, informational, and cultural arrangements. Spaces are always infused with technology, in the original sense of the word, drawing on its Greek root, *techne*, or the methods of craft. Practical action constitutes space, and the objects of contemporary technology—infrastructures of power, lighting, connectivity, and interaction—are merely the most recent elements that support that practical action.

In the end, then, the term "mobility" itself—and the images that it invokes—may be a source of trouble, misdirecting us and failing to draw attention to the important issues. We prefer to think not so much in terms of mobility as in terms of fluidity. When one talks of mobility, the ideas that are mobilized are discrete ones—ideas of this and that place, the places in between, and the stable objects that move from one to another. Mobility is thus dependent on fixed boundaries (those of places and objects) and fixed configurations (the relationships between places and objects,or among places, objects, and activities) among which we move. When one speaks of fluidity, a different set of metaphoric associations is made available. This set focuses on adaptation to continuous variability rather than selection among discrete options. By contrast, fluidity implies a continual reshaping or shape-shifting in changing circumstances; it suggests flows and connectedness as opposed to nodes and separation; it turns our attention to the ways that technologies and practices maintain an essential identity while adapting to local conditions; it reminds us that objects and activities might not only adapt to their changing environments but also exert a pressure on them and contribute to their shaping; and it notes that sometimes objects are dissolved by flows.

Marianne de Laet and Annemarie Mol (2000) employ the metaphor of fluid technology to rethink the fixity of analytic "networks" in science studies, particularly the commitments to clear-cut identities and fixed boundaries in the Actor-Network Theory (Callon 1986; Latour 2005). While our use of that metaphor here is directed toward quite different ends, we find it equally productive, for similar reasons. We are concerned with the ways that technologies are not simply deployed in spaces but also seep into them and, in doing so, transform and reshape those spaces. We are concerned, too, with the ways that space and technology mutually shape each other—especially when technologies act as lenses through which spaces are understood or provide new frameworks for representing space and for trading in representations. Technologies of all sorts—and certainly today, mobile and ubiquitous technologies—do not simply reflect a commitment to particular spatialities and cultural logics but are also involved in producing them. Infrastructures emerge where people gather, yet people also gather when infrastructures are provided—from public seating to public transit to wireless networks. Technologies provide us with ways to narrate space, to describe and articulate it, but narratives have a way of becoming self-fulfilling prophecies as accounts of everyday life become sedimented as understandings, expectations, and predictions.

These perspectives potentially open up new means for examining the relationships between technology and spatial practice. For instance, discussions of how location might be reported or made visible within ubicomp systems (Iachello et al. 2005a, 2005b; Smith et al. 2005) take on a new dimension when we look at the performative aspects of spatiality and its expressions. Cartography has always been a site for social and political expression (Wood 1992); the kinds of legibility that ubicomp technologies lend to space do much the same, although the habitual idea that values are expressed only in how technologies are used, and not in how they are designed, leaves this question unexamined.

Where might this leave us? We close this chapter with two possibilities.

First, if information technologies potentially provide lenses through which space might be read, then this might also offer opportunities for reflection on the different forms of representation that are captured there. What would be it to offer a technology that makes space legible in radically different ways? Again, we draw inspiration here from the alternate spatialities of indigenous Australians and Western settlers, since this is an ongoing site of difficulty. It is intriguing to speculate about a use of mobile technology designed particularly to allow people to navigate a space defined not according to Cartestian measures but instead relationally and historically—one in which people, along with their relationships and movements, determine the parameters of spatial experience. Imagine a handheld device presenting a map that highlights acceptable and unacceptable paths, centers of safety and danger, determined not in terms of the traditional map grid but rather in terms of the recent and anticipated movements of others and the social relationships that the device's owner has to them—the spatial exclusions that apply, forms of engagement that might condition movement, and expectations and implications of presence. Might this offer up a new way of understanding not just the space around us but the lived experience of being in space too?

Second, are there opportunities to create location-based technologies that eschew a commitment to the primacy of specific kinds of spatial representation? Any type of spatial technology operates in combination with a form of practice that lends meaning to particular kinds of movement, dwelling, and presence. In a study of Inuit integration of GPS technologies with traditional hunting and navigation techniques, anthropologists Claudio Aporta and Eric Higgs (2005) reject a superficial interpretation that technology displaces tradition and show instead the emergence of a more complex hybrid practice in which GPS takes its place alongside other forms of way-finding and spatial practice. What might it mean, then, to create location-based systems in which the commitment to a Cartesian representation was recognized as simply one among many—in which it existed alongside, rather than in support of, accounts of space in terms of biography, activity, and sociality?

By thinking in terms of fluidity and the mutual adaptations among technologies, places, and practices, we gain a different analytic perspective on the questions of mobility and most especially on its value. The values of mobility expressed in the vision of ubicomp go well beyond a computational environment that one carries around as one moves about in the world; its power lies instead in the kinds of flexible reconfigurations and fluid accommodations that characterize the different encounters between people, technology, and space. This opens up new avenues for design, certainly. More important, however, it exemplifies the reconfigured relationship between social and technological considerations in the way that we have been advocating throughout this book, not simply by placing one perspective in service of the other, but rather by highlighting their mutual constitution.

7 Rethinking Privacy

The Active Badge was a small, wearable device that transmitted a unique infrared signal every ten seconds to a network sensor, which could then accurately identify the transmissions and so pinpoint the device's location (Want et al. 1992). Initially conceived and deployed to help the receptionist at the Olivetti Research Lab (ORL) more accurately forward phone calls from the front desk to the call's recipient (there were no fixed lines into the lab), the indoor tracking system generated strong emotional responses from both those being badged and those using the location-based information. For the receptionist at ORL, the system was a boon (ibid., 97). For those being tracked, however, there were more complicated reactions: social fears, horror, concerns about violations of personal freedom and individual privacy rights, and worry over the sanctity of break rooms and bathrooms (ibid., 99–101). For the original inventors, Active Badge was always intended as an extension of other forms of office systems and management, like phone systems, fire alarms, security, and climate control. Yet between 1989 and 1992, as the system was deployed in universities and industry research centers in Europe and the United States (AT&T Laboratories Cambridge 2002), further debates about the merits of tracking, mechanisms for thwarting the system, and the politics and polemics of surveillance ensued (Harper 1992; Agre 1995).

Dourish was working at EuroPARC when Active Badge was deployed there in 1990. He was one of the original "refuseniks" who did not wear the Active Badge at all times and was a frequent critic of the system. As the third deployment site, EuroPARC represented a different social, organizational, and research environment: researchers spent more time in their offices, a departure from the practices at ORL where hardware laboratories were frequent retreats; the researchers were a far more heterogeneous blend of social scientists, computer scientists, and engineers compared to ORL's strong engineering tradition; and there was no need for transfering calls because each researcher had a separate phone line and number.

When the Active Badge was utilized at EuroPARC, it was incorporated into several research projects, including Pepys (Newman, Eldridge, and Lamming 1991). Pepys automatically generated a diary-like account of activity based on where the badged researchers had been physically located and with whom they had been colocated. Pepys was focused on meetings and episodes that involved two or more of the EuroPARC researchers. Choosing not to wear one's badge, then, created problems for this project and the data it produced. Nonparticipation in the system affected not only the nondisclosure of one's own location or context but could actively fragment the diaries of others, rendering them less accurate. For the EuroPARC researchers, compliance now had an additional moral and ethical layer—it was less a personal choice and more a matter of scholarly responsibility.

In ubicomp's narrative arc, the Active Badge deployments have a prominent place. It was one of the first attempts to build a functional sensor network with a working application. It was also one of the first instantiations of smart-location technologies, foreshadowing the motes and sensor work that would take place at the University of California, Berkeley and other sites. And perhaps most importantly, it was read as a sign of things that could be possible when computational technology got off the desktop and made its way into more personal forms and formats. The Active Badge has set forth and privileged the linking of ubicomp technologies and privacy concerns. As a deployment, and in its various recapitulations as conference papers and presentations, it set the tone and framed the debate for years to come. Indeed, we might argue that it has become the template for discussions about privacy and ubicomp. Ubicomp deployments will always engender privacy concerns, so the line of reasoning goes, but they can be addressed and overcome—the technology is neutral, so the misuses are abuses of the system not of the system's marking. As Adam Greenfield (2006, 19) writes describing the Active Badge project, "Original sin came early to ubicomp."

It is impossible to contemplate authoring a book engaged with ubicomp that did not address privacy for, although privacy was not originally part of Weiser's definitive agenda-setting article, it has been interpolated with the ubicomp conversation almost ever since. But in this story, privacy is almost never defined; there is an operating assumption that we all have a shared understanding of this term and its various dimensions—that it is stable and universally held. Dimensions of gender, race, class, and age are rarely if ever discussed, and likewise for cultural or sociopolitical forces, power, or regimes of surveillance and control. Privacy is sometimes placed in opposition to public in the dichotomous framing of public/private—although in this framing, public is similarly almost never defined. Today, despite this lack of nuance, or perhaps precisely because of it, privacy is constituted as a priori in ubicomp; it is treated as the dominant sociotechnical concern.

The rise of electronic information and mobile technologies has been accompanied by an increase in sociotechnical privacy concerns, whether prompted by the large-scale loss of personal data by government or commercial entities or by the threats of GPS-equipped mobile phones for personal tracking. Nevertheless, despite widespread concern about privacy, good conceptual models for understanding what is at stake are lacking. In part, this is because the very concept of privacy is embedded in a specific series of cultural and legislative frames, suggesting that perhaps it is not the best or most productive place to start. It is impossible, we contend, to talk coherently about privacy without also talking about trust, secrecy, risk, danger, lies, control, security, identity, morality, and power.

In this chapter, we attempt to significantly complicate the notion of privacy. We recontextualize the discussions of privacy in, and of, information systems by using the related concepts of secrets and lies to examine how people interpret, value, and understand flows and exchanges of information. How do we want to think about the ways in which people report location to others, for example? How does an examination of the maintenance and enactment of social relations in everyday life help us reevaluate the technological notion that "information wants to be free"—a notion that looks quite different when we think about Western scientific data, axial religious practices, or even the secret sharing of North American teens? In line with the themes that we have been developing throughout this book, what emerges is a focus on the practices of articulating and sharing information as a means of cultural production, a way in which people engage in meaningful social interaction and the negotiation of collective meaning.

"Are You Menstruating?": Cultures of Privacy

Bell (2005, 2006b) was in Beijing in 1999 with a colleague conducting her first ethnographic research in the People's Republic of China. One time, when Bell and her Chinese colleague were at a restaurant, a waiter appeared to take the tea orders. Bell recalls: "Yuet and I were talking about the next day's work, and the tea man appeared, in this remarkable yellow and red costume. He and Yuet had a conversation, and Yuet looked at me and said, 'Are you menstruating, the tea man needs to know.'" As it turns out, the position in one's menstrual cycle helps guide choices about food and beverage selections. Had Bell been menstruating, there was a special house tea with blackened chicken bones and a particular set of medicinal plants. Bell has had similar conversations in rural Italy, where the size of bubbles and mineral mix in water have healthful properties. In both instances, however, information that is regarded as private in a U.S. context was fair game in public spaces with comparative strangers.

We argue here that privacy is differently constituted—not simply differently implemented—in different cultures. That is, it is not just that different people have different ideas about what things should be private but rather that around the world there are clearly different ways in which privacy is mobilized, made sense of, and managed. Some of these differences are strictly cultural or religious, some are driven by geopolitical concerns, and some are motivated by different notions of the relationship between the nation-state and its citizenry. In Korea, for instance, the home is an intensely private space, restricted to close family members. There are further gradations of privacy within the home, with the *anbang* (or women's room) strictly off-limits to any visitor and sometimes restricted just to the women and children within the home, with the entrance frequently hidden by a decorative screen (Seo 2006; Yoon 1979). Technology moves freely through and across these borders, and mobile phones, which have a ubiquity in Korea achieved almost nowhere else, ring in every room. Egyptian homes have similar gradations of access. Beautifully decorated and furnished salons are open to visitors and family, but there are more private spaces in back where televisions and computers keep company with the family and where women can shed their head scarves and protective clothing, forgoing modesty for family comfort (El Guindi 1999; Hasbrouck and Faulkner 2006).

In India, by contrast, most homes are incredibly open spaces. Extended family, close friends, and neighbors make regular and unannounced appearances, spending hours socializing throughout the house (Asokan 2008). In Indian culture and Hindu cosmology, homes anchor the purity/pollution dichotomy, with the polluted sphere encompassing public spaces along with the messiness of activities and actions therein (Dumont 1966/1980). A similar sense of permeable domestic spaces has been noted in a study of transnational migration and technology usage in Thai households (Williams, Dourish, and Anderson 2008). In this conceptualization, purity and privacy are not equivalencies, and in fact different kinds of practices arise as a result. During Bell's fieldwork, she encountered men consuming pornography in India's vast network of public cybercafes because to do so at home would be to violate one's home, insult one's family, and demean one's wife. In Korea, Egypt, and India, notions of privacy are important, but they have profoundly different meanings, histories, and interpretations. The different manifestations of privacy impact everything—from social relationships to divisions of space and labor, from access to information to the movements of goods, services, and people—and thus complicate privacy as it is encountered in ubicomp.

Privacy as it is deployed within the ubicomp tradition, however, rests on distinctly Anglo-American cultural underpinnings. It is intertwined with a number of other key Western narratives and social constructs. Principal among these interconnections is the post-Enlightenment notion that the individual is the smallest indivisible unit of personhood (Hsu 1953, 1983; Schneider 1980). This means that in the West privacy is often tied to the individual. Thus, the individual has the right to control over a body of information/behavior/physical objects at the expense and exclusion of all others. Cutting across these notions of individual control is also the belief that keeping things private (i.e., strictly to oneself) is a form of social good and social entitlement. Paradoxically, there is also a strong belief that certain kinds of information should exist in the public domain and be open to everyone, and that once things are said they cannot be unsaid or taken back into the private realm. Many of these conceptions are rooted in the foundational documents, ideological statements, and judico-political actions of the last two-hundred-plus years of U.S. history.[1]

As we have suggested earlier in this chapter and in chapter 3, not all cultures around the world operate within this same cultural logic. In many cultures, the individual is not the smallest unit of social (or even legal) identity and is not the entity to which privacy defaults; instead families, households, clans, lineages, descent groups, temples, or villages might be the social unit of preference. As such, where we might expect that certain kinds of information—especially around health, money, politics, and sexuality, among others—be kept private to ourselves, in some other cultures that kind of information might be fiercely guarded within

1. One could even suggest, somewhat radically, that there are certainly schools of thought (most in the political economy side of the social sciences) that would assert that privacy is a post-Enlightenment, late-capitalism ideal; some might even argue that it is a form of false consciousness. After all, how much privacy do Americans really have? Indeed, even the privacy debate in the United States is complicated and hardly stable.

an extended family or community. Clearly other cultures are drawing the lines about what is public and private quite differently. Try to imagine a barista at a Starbucks in Cincinnati inquiring as to one's menstrual cycle and offering particular lattes to those who are ovulating or menopausal; or perhaps imagine all the televisions at a Best Buy, Currys, or Harvey Norman broadcasting hard-core pornography instead of football. We would conjecture that this would be shocking. Bell's fieldwork in Indonesia revealed families who shared a single email address and a basket of phones even when it was possible (economically and otherwise) to have one each—because sharing was seen as an important cultural value.

Furthermore, it might well be the case that while privacy is a dominant sociotechnical concern in the United States with regard to the Internet, it might not feature as the first set of social concerns or responses in other cultures and countries. There might well be a range of other reactions, equally significant, just following different sets of cultural scripts and logics. In China, the conversation is much more framed by ideas of state surveillance as well as notions of threats to national and community stability—for better or worse. In the Aboriginal societies in which Bell grew up, people worried about secrets and stories that were considered sacred (as opposed to profane), and privacy just wasn't a topic that made sense; even as people engaged with the Western nation-state, their concerns continued to be framed by worries over secret knowledge, not privacy. We were even struck by a recent story about the new Wellcome Wing at the British Science Museum, which installed biometric fingerprint readers so that museum patrons can track their own experiments from one installation to the next. As opposed to a concern with where those fingerprint scans are going and who has access to them, the principal fear seemed to be about hygiene and the spread of disease. All of this suggests that privacy and security are hardly stable transnational phenomena worldwide. Sometimes privacy might not be the right question or design imperative, and likewise for security. After all, not only are these cultural constructs; they are not always the right ones.

Making Sense of Privacy

So how might we make sense of privacy in a way that can be productive for ubicomp, moving the discourses and practices forward? It is clear that while privacy is a long-standing part of the ubicomp research agenda, the issues remain largely intractable. So where might we look for different theoretical and methodological constructs and insights? How might we account for privacy as a flexible and fluid concept, rather than one that is fixed and unmarked?

We might usefully start by identifying a series of different ways that the topic of privacy is approached in the research literature (Dourish and Anderson 2006).

The first and most prevalent way of dealing with privacy in the technical literature is to adopt an economic metaphor. The key idea here is that privacy is bound up in a trade-off between risk and reward, or the cost and benefit associated with sharing, revealing, or transmitting information. The notion that information is an object is a metaphoric foundation for much discourse in information and computer science (Buckland 1991); here, this idea is extended so that information is modeled as a commodity that can be traded (Patil and Lai 2005). This allows the apparatus of microeconomics to be brought to bear. The economic approach to privacy then models collective action as the outcome of individual decision making by rational actors optimizing for individual benefit. The economic model is at the heart of many proposals for interactive and ubicomp systems, including Place Lab (Schilit et al. 2003; Hong et al. 2003), recent anonymous location-based systems (Gruteser and Grunwald 2003), and the use of RFID devices in physical stores (Floerkemeier, Schneider, and Langheinrich 2004).

As a sole explanation for privacy practices, the economic has a number of problems, conceptually and hence as a model for design. First, studies of actual practice fail to display the sort of rational trade-off that this model would suggest (Spiekermann, Grossklags, and Berendt 2001). Recent research in the area of behavioral economics suggests that traditional rational-actor approaches fail to adequately account for everyday behavior even within their own terms of reference (Rabin 1998; Kahneman and Tversky 1979.) More problematically, though, economic models fail to recognize that privacy is essentially a social practice. A trade-off or exchange between rational actors surely fails to capture the sharing of intimate secrets between lovers (Richardson 1988) or the morality of full disclosure in closed groups (Kleinman and Fine 1979). Economic models may provide a gloss or explanatory account of information practices, but accounts and motivations must be distinguished (Schutz 1943). While it is possible to incorporate some of these elements into economic discourse (e.g., in discussions of social and cultural capital), complementary models can help us look at privacy as a part of everyday life.

The second approach is to think of security as a practical phenomenon. This turns attention away from abstract information exchanges and toward the practical detail of what people do. Security, in this argument, is not an abstract feature of ideal settings; it is a practical, ad hoc, in-the-moment accomplishment of social actors achieved through their concerted actions in actual settings. When we take this perspective, a quite different set of questions emerge. How do people go about doing work securely? How is the difference between public and private demonstrated in the ways that they go about their business? How are private matters organized and accountably produced (Garfinkel 1967)?

The focus on practice reconstitutes privacy as a pervasive element of everyday settings, which extend beyond the boundaries of any or all computer systems and incorporate organizational arrangements and practices, the physical environment, and so on; it draws our attention to how people orient toward statements and actions so as to reproduce the cultural logic of privacy, boundaries, and social groupings.

A third approach to privacy and security is as a discursive phenomenon. Language does not simply describe the world but also is part of the process of constituting and shaping the world that we experience. So the issue here is to understand how the notion of privacy and security are used to categorize activities, events, and settings, separating acceptable (secure) actions from unacceptable (insecure) ones.

Clearly, any such use of language embodies a particular perspective: Security of what, for whom, and from what? What risks are implied? And who gets to define them? For instance, much discussion of information security occurs in corporate contexts, and corporate security directives typically place organizational conveniences ahead of personal ones. The determination of what information security means is one that not only condones some behaviors while outlawing others but reflects power differentials within the organization too.

These three vantage points complement each other and reveal different aspects of the settings within which privacy and also security are managed. Largely, though, it is the first view that has held sway in information systems design; mechanisms such as access control, for example, are designed as encodings of exchange value. The latter two alternative approaches share a stress on information practices as collective rather than individual phenomena, and so they turn our attention in different directions. In particular, this third approach to privacy and security as a discursive phenomenon places primary emphasis on the broader social and cultural logics of security—the contexts that shape the distinctions between secure and insecure. When looking at these contexts, it becomes apparent that privacy and security cannot be analyzed independently but instead must be considered alongside such related concerns as risk, danger, secrecy, trust, morality, power, identity, and so on.

Looking outside the ubicomp community reveals other approaches that are useful. The privacy debate looks different in other disciplines and locations; how it is cast and recast under different governmental, economic, and social regimes all come into play, and it is often not accorded the kind of centrality it finds as a social concern within ubicomp narratives. In fact, privacy is not a traditional object of anthropological or even ethnographic inquiry. While there are a few notable exceptions (e.g., Hsu 1963), anthropologists have historically been more inclined to examine the public/private dichotomy as it relates to space and social relationships (Dumont 1966/1980; Douglas 1966) and the secret/sacred (or sacred/profane) dichotomy as it relates to information and knowledge (Lévi-Strauss 1970; Durkheim 1912/1965). These examinations offer a variety of different ways of approaching, situating, and decoding privacy—information is not, for instance, democratized but instead is accessed only as one acquires ritual power or undergoes increasingly rigorous initiations (Herdt 1994; Meggitt 1962; Van Gennep 1960). For Warlpiri men, subincision is traditionally a marker of adulthood and also signals a certain kind of social authority and standing, with access to secret knowledge (Meggitt 1962). In this culture, some information remains immutably classified as secret—its currency comes from noncirculation. (Imagine what such a formulation of information and knowledge, and access thereto, might look like in ubicomp—initiation ceremonies, body markings, or rigorous testing before one could, for instance, participate in Active Badge.) Contemporary ethnographic accounts of resistance, surveillance, and the gaze, especially as influenced by Foucauldian notions of power and the Panopticon, represent another means by which privacy has been encoded and decoded within the literature (Foucault 1975). Here privacy is linked to larger sets of social and regulatory forces and is understood as an evolving set of cultural inclinations. As such, the constellation of practices that might be glossed as privacy concerns is large.

We suggest three further ways of reapproaching the problematic of privacy for ubicomp: through privacy dichotomies, lying as a privacy practice, and finally the clustering of privacy with an emerging set of sociotechnical concerns.

Privacy Dichotomies

Much as ubicomp has treated privacy as a stable and fixed object, it has also been regarded as part of a fixed and knowable dichotomy or opposition: private/public. Privacy is often understood or framed by that which it is not—that is, in the open, in plain view, or public. Yet public is not privacy's only oppositional or dichotomous frame of reference. Recent work within anthropology, sociology, and design all point to different sorts of dichotomies, oppositions, and spectrums.

Building on her early work on the daily negotiations between home and work life, Christena Nippert-Eng (1996) offers a nuanced and thoughtful accounting of privacy practices in U.S. daily life. Based on ethnographic fieldwork with middle- and upper-middle-class Americans in 2001–2002, she writes that for her research participants, "privacy largely centers on the degree to which an individual believes she or he has control over the accessibility of things that are 'private.' This might be some aspect of self (including one's body), a thought, a behavior, relationship, piece of information, chunk of time, a certain space, or an object" (Nippert-Eng 2006, 1). This privacy story is hardly a simple or straightforward one—privacy here, and one supposes privacy concerns, extend from the body to time, space, social relationships, and even knowledge and information. As Nippert-Eng (2007, 12) rightly points out, this means that addressing privacy concerns in the United States involves accommodating a surprisingly wide range of issues and stakeholders. Her work uses the more familiar private/public dichotomy but also a framing of privacy/publicity (Nippert-Eng 2005a, 2005b, 2006, 2007; Nippert-Eng and Melican 2004; Nippert-Eng et al. 2005). This second formulation turns out to be a more analytically useful one; here privacy is understood as the condition of pure inaccessibility, and publicity as the condition of pure accessibility, while a spectrum of activity and understanding lies between them (Nippert Eng 2007, 3). This notion of publicity and the accessibility on which it is based is a helpful way of rethinking public: What spaces, practices, and technologies might fit into this space? How might thinking about publicity concerns for ubicomp change our perspective?

Based on ethnographic and design research in India over the last three years, Ashwini Asokan (2008) offers a different formulation of privacy. In her interviews, the Western formulation of privacy was viewed quite negatively in middle-class Indian homes. She writes that privacy was variously described as "something wrong, as isolating, rude, anti-social, selfish, taking time away from the family and duties and as something required when one needs to hide something or is doing something against commonly accepted rules" (ibid., 12). She recounts instead the ways in which Indian households mobilized a different form of privacy, conceptualized vis-à-vis shared objects, spaces, and social networks in the home. Here "shared" becomes privacy's cultural pair, not "public." Asokan's work further complicates notions of the shared in two ways. First, she suggests that while shared/private are oppositionally linked, they are not always mutually exclusive. Second, as she is quick to point out, the use of "our," as in "our space," "our bed," or "our television," does not actually mean fully equal, democratic, distributed ownership and use. Instead the sharing evoked here, within the context of Indian middle-class homes, also encodes a complex and ongoing socially sanctioned negotiation of familial hierarchies, prioritized activities, and individual actors. Echoing work done elsewhere (Ito and Okabe 2005a; March and Fleuriot 2006), Asokan (2008, 2) also observes that "Indians have developed sophisticated ways to make themselves visible or invisible depending on the contexts of their surroundings," especially in their homes. What might be understood as privacy in the ubicomp research agenda is more likely to be achieved outside the home in what we might regard as public places—cafés, theaters, cyberarcades, parks, and malls. This notion of creating privacy in shared domestic spaces or openly nondomestic places, of being private in plain sight, raises interesting tensions for ubicomp; such levels of contextual and locational complexity is not accounted for in most smart environments.

Lying as a Privacy Practice

Another possible approach to reformulating privacy comes from a closer examination of current digital-intermediated privacy practices: How are people engaging with online spaces, applications, and experiences so as to ensure privacy? How are people utilizing new information and communication technologies to ensure (or perhaps violate) privacy? Of the range of privacy practices engaged, perhaps the most interesting centers on the use of providing false information as a protective privacy strategy. Specifically, we are interested in how we might utilize the notion of digital deception and lying to help reframe the conversation about privacy.

People lie on any number of websites and services about their demographic particulars: age, gender, and date of birth. Some of these lies are necessary for participation; for instance, a disproportionate percentage of MySpace users are over one hundred—a demographic anomaly more likely driven by U.S. regulations setting a minimum age for participation than some flourishing of senior-citizen digital activity (boyd 2007). There are also lies that seem related to a different kind of privacy—where the concealment of one's identifiers might change the outcome in an undesirable direction. For example, researchers at Cornell University have found that 100 percent of those participating in online dating lie about something;

men lie about their height, systematically overestimating it by an average of two to three inches, and women lie about their weight, undercalling it by on average three to five pounds (Hancock 2007; Hancock, Tomas, and Ellison 2007; Hancock et al. 2008). While lying here might be a form of deception, it is arguably also a form of privacy. Writing in early 2009, one might even assert that microblogging systems like Twitter, ringing with detailed descriptions of mundane activities, concentrate on the charms of small acts of confabulation rather than social history in the making. Amid all the publicity, to borrow Nippert-Eng's framing, what might actually be left unsaid? What might be hiding in plain sight on Twitter and other sites like it? Fascinatingly, recent research suggests that online deception appears to be an enjoyable activity—for those engaged in acts of digital deception, guilt, fear, and shame are largely absent (Caspi and Gorsky 2006). The implications for this on privacy practices in the ubicomp space are fascinating to contemplate.

For as much as people are lying about their aspirations, identities, and intents, they are also using various computational devices to lie about their locations and contexts. According to one recent report, some 45 percent of British cell phone users report lying about their whereabouts in calls and texts (Sturgeon 2004), and that includes only those who are telling the truth about lying. There are also more deliberative services that allow individuals to create alternate truths about where they have been and what they have been up to (Richtel 2004). James Katz, a communication scholar, has described these practices as "practically an arms race between the technology used to locate people and track behavior—global positioning systems, for instance, and caller ID on phones—and technologies intended to deflect surveillance, like audio for fake background noises" (quoted in ibid.). His claim is that the rise in location-tracking software, personal spyware, and the like, has been countered by alibi services, anonymization technology, and outright lies—the escalation of services and technologies designed to catch liars, or to perpetuate the falsehoods, is an interesting one. And it is complicated in some ways by the fact that newer generations of technologies have an even greater potential to tell the "truth" unbidden and thus potentially violate privacy practices. The increasing availability of location-aware devices (e.g., GPS) and environments, context-aware devices (sensors and motes) and algorithms (triangulating sources), and the pervasive nature of data trails/search records created through Internet activity and by mobile devices (some countries allow the sale of data trails) suggests an abundance of ways in which privacy could be violated and where the current practices of lying to protect privacy might be less than optimal. Some governments, regulators, companies, and researchers are grappling with these issues with levels of granularity of location information, location anonymization, and imprecision techniques, but these might have other unintended consequences, and if Katz is right, they could generate a new set of disclosures.

At the same time, as we acknowledge the critical roles of prevarication and falsehood as critical components of contemporary (and, we imagine, future) information practice, it is also important to recognize a second lesson here: not everything that looks like a lie *is* one. Information systems—and perhaps especially ubicomp systems that rely on sensor systems to instrument the everyday environment—have a tendency to place within their representations a notion of "ground truth," from which all deviations seem problematic. Yet as we see with cases of spatial location, not all deviations from sensor reports are lies. Reporting "I'm on my way" in response to a query while sitting solidly on a chair in front of a computer is not necessarily a lie; it may indeed be an entirely appropriate response, and a thoroughly truthful one at that. Spatial reference is ambiguous, and its production is highly context dependent (Schegloff 1972; Curry 2002). The attempt to compare human action to sensor data in order to produce a computationally tractable distinction between truth and falsehood seems reminiscent of dystopian science fiction, and certainly a vivid demonstration of the mismatch between binary logics and everyday practice. This is not just an argument for "shades of gray" but rather an observation that lying and truth telling are social practices, not easily understood or distinguished through sensor networks. So as well as highlighting the complexity of social practice, an examination of lies in information systems also reorients us from the representation to the production of social action and its context (Dourish 2004).

Clustering Privacy with Other Sociotechnical Concerns

Privacy is often regarded as the signal sociotechnical concern within ubicomp, but recent research suggests there is a clustering of privacy with an emerging set of sociotechnical issues. Over the past decade, computers and the Internet have been linked to a well-rehearsed set of sociotechnical concerns, including privacy, trust, security and risk of identity loss, access to inappropriate content, threats to children, and regulation. As the Internet has moved beyond the traditional computer platform, and as new connected devices, infrastructures, and services have allowed new experiences, it has given rise to a range of additional sociotechnical concerns. These include issues of reputation and image, reliability, access, participation, physical health and well-being, environmental sustainability, responsibility, authenticity, authorship, ownership, surveillance/control, and a whole host of anxieties around cultural health, including fears about digital literacy, dumbing down, and a loss of cultural distinctiveness. Of this increasingly large cluster of sociotechnical anxieties, the ones most linked to the more traditional privacy fears seem to include concerns regarding trust, reputation, and image. Recent work by researchers in Bell's organization at Intel suggests that privacy clusters especially tightly with this particular range of other sociotechnical concerns as computational awareness and intelligence move into the realm of consumer electronics.

In 2008, Intel researchers Jay Melican and Sue Faulkner conducted research in Spain, South Korea, and the United States, focusing on the ways in which individuals understood the trade-offs between making personal data more readily available to their service providers in order to gain customized experiences and their own sense of privacy. This research, not yet published, revealed interesting tensions, strong cross-cultural differences, and a gradation of information that could be revealed without a sense that privacy had been violated. Their findings are noteworthy for the kinds of information that were regarded as most private and those that were more willingly disclosed. Nippert-Eng (2007) has written about the notion of selective concealment and disclosure in U.S. middle-class homes. Melican and Faulkner also find that what was to be kept private was situational and contextually dependent.

They further contend that there were roughly defined concentric rings of increasingly classified information. At the outermost edge of this was information about content practices and preferences (i.e., television shows, time of viewing, etc.); next in terms of privacy came basic profiling information (i.e., age, gender, and address); then consumption patterns and behaviors (i.e., recent purchases); and next were online patterns and behaviors (i.e., search and browser histories). The two sources of information considered to be most private in this study were socioeconomic information (i.e., income, levels of education, and renting or owning homes) and, at the center of this circle, friends and family. In this layering, unexpected information is contained within the bounds of what is private: the kinds of content that is consumed, when and how, and the nature of one's social networks. In both instances, the concern here is around reputation, trust, and identity politics. There was a clear sense running through the data that Melican and Faulkner collected that protecting one's friends, family, and work colleagues was an important part of one's own privacy practices; releasing data that affect friends, family, and work colleagues was unacceptable, and information that might impact them was considered nonsharable (i.e., mobile phone, instant messaging and email activity, and social networking participation)—it was private. Maintaining good privacy practices, then, meant keeping others safe too.

Risk, Danger, and Morality

Although there is an extensive literature on risk, and in particular on social theories of risk, this rarely features as part of the ubicomp discussions of privacy and security technologies. Any notion of security, however, must implicitly or explicitly turn on questions of risk and danger, especially those risks against which we must be secure. To formulate a state as secure is to define what dangers are faced. Similarly, defining specific states or activities as insecure marks them as inappropriate and sanctionable.

What this suggests, then, is that the ways in which risks are formulated and used to define states of security are ways in which socially acceptable and unacceptable behavior is demarcated. It is thus valuable to pay attention to the sources of these attributions.

Of interest here, security is defined with respect to a set of perceived risks. Mary Douglas and Aaron Wildavsky (1982) explore the cultural aspects of risk formulation and selection. They make two primary observations. First, they note that the selection of particular activities and objects as risky and matters of concern, and the passing over of other activities and objects as not worthy of being labeled risky, is not a purely rational or objective process but rather reflects cultural, political, and moral judgments. Steven Shapin's comments (2003, 2004) about the morality of diet are a nice case in point; pointing to the peculiar symbolism of food, he notes a transformation in how the risks of diet have been discussed, from a historical model in which obesity was the morally reprehensible consequence of gluttony and moderation was to be exalted, to present-day diet books that offer the ability to "eat as much as you like," moving the responsibility for obesity from a person's willpower to a body's metabolism or foodstuffs themselves (e.g., "bad carbs"). The source of risk has been transformed, and in so doing, the risk has been changed from a moral and spiritual one to a chemical and physiological one, in line with changing cultural attitudes. This is not an isolated example; we can see this all around us. Consider the role of U.S. cultural attitudes toward individual mobility, justice, and technological progress in debates around the risks of transportation, the death penalty, or nuclear power.

The second major issue that Douglas and Wildavsky examine is how risk perception should be read relative to social structure. If we read risks as potentially endangering not individuals but rather social structures and "cultural truths," their second observation takes this further to suggest that different social collectives will have different interpretations of risk depending on their position relative to the social structures that might be in question. They spend some time looking at alternative perceptions of risk by those who are placed in more or less central and stable social positions. This is reminiscent of Brian Wynne's exploration (1992) of scientists' and farmers' relative knowledge of nuclear technologies and agricultural practices in discussions about the impact of the Chernobyl disaster. Wynne's study points to the different interpretations of risk between those in more central and marginal societal positions as well as pointing more generally to the tension between "inside" and "outside" knowledge when epistemic communities encounter each other. For these different epistemic communities, the rhetoric of risk may take quite different forms. For example, the rhetoric of risk is on display in debates over finding sites for storage facilities for low-level radioactive waste. Louise Bedsworth, Micah Lowenthal, and William Kastenberg (2004) highlight not only the way in which those who opposed the development of new facilities brought up the potential risks to the ecosystem and local residents but also how those who were more invested in scientific accounts of the facility's safety indicated the risks to those who stood to benefit from the activities that might generate the waste, such as cancer patients suffering as a result of the constraints on scientific and medical research into treatment strategies. The debates around risk become, in this case, struggles over the rhetorical strategies by which risks are defined; in turn, these strategies are associated with different epistemic communities.

What is important to note here is how definitions of risk mark distinctions between acceptable and unacceptable behavior and, at the same time, relationships between different social groups. In her research, environmental psychologist Kristen Day (1995, 2001) has looked at how men's attitudes toward women's vulnerability in public space are manifested. She observes not just the different perceptions of women's vulnerability but also how they act as a site to reinforce and reproduce cultural logics of action and interaction. As Day (2001) maintains, these cultural logics of risk also serve to mark regions of the environment as off-limits and act as a form of social control for the "at-risk" group. The very fact of risk assessment, indeed, is a means by which a socially meaningful designation of "at risk" can be formulated, in much the same way as disease risk factors can create a new social category of the "presymptomatic ill" (Lock 1998; Fosket 2004; Parthasarathy 2004).

In the case of information system security, these same considerations are at work (Weirich and Sasse 2001; Dourish et al. 2004); this is hardly a surprise, given the basic relationship between security and social structure that we have been sketching. Researchers in privacy and security recognize, of course, that assessments of risk are highly variable and relative. Yet our point here is that such assessments are collective rather than individual phenomena. The significance of this observation for technological settings revolves around the enacted aspects of social practice—that is, the social meaning of the distinctions being drawn is not simply a feature of the natural world but instead is continually reproduced in everyday social behavior. Since security tends to feature in design deliberations as a natural fact rather than a social accomplishment, alternative formulations of security and privacy, and the emergence of norms of action and interpretation as a consequence of social engagement, are typically erased. If we accept that privacy and security are formulated relative to normative conventions of risk and danger, and that, in turn, these both give rise to and are shaped in everyday social action, then the design question becomes not how we can reflect social norms within information systems but how we can reconfigure information systems as sites for the production of social and cultural values.

Sociologist Ulrich Beck (1992) has suggested perhaps the most extensive relationship between risk and social structure. Beck's "risk society" is an outcome of processes of what he, along with Anthony Giddens and Scott Lash, has called "reflexive modernization" (Beck, Giddens, and Lash 1994)—a form of modernity that arises when society becomes conscious of its ability to radically transform the circumstances of its own existence, through science, technology, and the forms of organizing available to technological states. The risk society, in Beck's terms, is therefore one in which dangers and hazards are continually in focus. As he points out, risk is not uniformly distributed, suggesting that the distribution of risks—and its relation to knowledge, resources, and power—may play as much of a role for social theorizing in the risk society as the distribution of wealth has traditionally done. From this perspective, then, figuring risks and the identification of dangers are as much a product of particular forms of social life as they are consequences and drivers of technological practice. Thus, particular forms of those risks and dangers—and the mitigations that might be adopted—need to be seen as products of social life as much as of the natural world.

Secrecy, Trust, and Identity

A useful way to rethink the question of privacy is by the closely related but differently valenced notion of secrecy. It is impossible to talk about the keeping and sharing of secrets without talking about those groups among which secrets are shared or from which secrets are kept. Secrecy, identity, and affiliation are intimately related. Secrets express intimacies and mark groupings, dividing the world into "us" and "them"—friends, families, fraternities, and more. Indeed, the common feature of studies of secrecy is the way in which the practices of keeping and sharing secrets are ways in which affiliation and membership are managed as well as demonstrated. Again, the use of information to demarcate boundaries is no surprise when we look at the sociological literature; what is important here is how contemporary computational practice can be illuminated by these views.

There are two levels on which we can explore the relationship between secrecy and collective identity. First, on a relatively superficial level, secrecy is intimately connected with social boundaries. Secrets—shared information whose disclosure would somehow endanger the involved parties—simultaneously cement a bond between those who share them while marking their differences from those with whom the secrets are not to be shared, and this operates across a range of settings from secret societies (Erickson 1981) to illicit relationships (Richardson 1988). A second approach is more relevant to our interests, though, which is to examine the notion of "cultures of secrecy"—social settings that give meaning to certain kinds of information, denoting them as secrets and hence giving meaning to patterns of information sharing. We note two aspects here.

Cultures of secrecy make information and its flow meaningful to those who are part of them, and the flow of information itself serves to reproduce those cultures. Don Merten (1999) introduces this concept in an analysis of information practices in high schools. Unsurprisingly, secrets are used to reinforce and rebel against authority relationships between children and parents, teachers, and adults, but they are also a crucial resource in managing and navigating the complex world of peer social relations. Secrecy itself, as a marker of a social relationship, is frequently in these cases more important than the content of the information; secrets may be used strategically to cement alliances and deepen friendships. Of particular concern is not just the fact of secret sharing as a cultural marker of intimacy but also the process by which people learn how to share, keep, and use secrets, and how the dynamics of peer and family relations are sites for the negotiation of norms about what is to be shared and under what circumstances. Part of the process of keeping a secret is recognizing one in the first place, which requires sensitivity not only to the information itself but to the costs of disclosure as well (which may themselves lie largely in risks to other social relationships). What is significant here are not the secrets themselves but the collective orientation toward practices of secrecy.

Indeed, knowing how to treat information—as sensitive or not—becomes a sign of membership and competent practice. In an ethnographic study of amateur mushroom enthusiasts (Fine and Holyfield 1996), trust and secrecy both play an important role in the development of group cohesion. As new members join the group, they must learn to trust in others' identification of edible and inedible mushrooms and then use of them in various dishes produced for general consumption; at the same time, they must learn the group's conventions and practices toward members' knowledge of particular mushroom-collecting spots (which are highly personal and carefully guarded). Asking someone for their favorite spots (or rather, expecting to be told someone's favorite spots) is highly inappropriate. Members will go to lengths at times to avoid being heard as requesting this information. As part of the process of enculturation, new members must learn what sort of information is to be shared and what is not, and must develop new understandings of the norms that govern information use.

In the area of information systems, some studies indicate that this may be a fruitful line of inquiry (Anderson and Dourish 2005). In a study of long-haul truckers, similar practices of trust and secrecy proved key in the group's social life. As people become truckers, they learn from other truckers the secrets of running with heavy loads, driving longer hours, and navigating effectively and cheaply with wide loads. They also learn culturally appropriate control over information flows, such as temporarily disabling GPS cab monitors or respecting conventions about known secrets. For example, it is extremely inappropriate to ask a trucker for a contact in a city to get a return load or a password to a website that provides return loads. In truck stops where a trucker can use a laptop, other truckers coming into the space go to great lengths to avoid looking at the screen and probe to be sure that the trucker isn't working to find a load, in case they seem to be violating this convention. Truckers' collective, normative information practices define and mark group membership.

These practices may also manifest themselves in "not noticing" information in the first place, as in a study of seniors in an assisted care facility (ibid.). The facility had attached load sensors to the seniors' beds to track their weight for health reasons. Yet the information so gathered and reported to family members could also indicate that their parent was sleeping with someone (and through the use of RFID, could potentially specify who). Family members were upset with the management for telling them this information, so it was not reported again and became invisible information. This was in essence both a collective decision and one that reflected the power dynamics at work (since the family members pay the bills and so hold ultimate power). In fact, this constituted the reemergence of a previously agreed-on information practice of the old community. The service people on hand in the house had always known this type of information but chose "not to see it" unless they were explicitly asked. They felt it was not their job to do so. It was agreed-upon invisible information, or a safe behavior in that context.

Appropriate information system design, then, recognizes that information practices—the selective sharing of information along with its appropriate management, including forgetting and not noticing—are not only embedded within social groups but also serve as ways that the distinctiveness and boundaries of groups may be identified and reinforced. Appropriate hiding and sharing of information is a marker of social affiliation and a way that membership is accountably demonstrated. Objects and artifacts are not inherently public or private; rather, as information is strategically deployed to shore up or break down boundaries between people and social groups, these categories are negotiated in use (Nippert-Eng and Melican 2004). Information technologies provide new ways to turn identity into an actively managed component of social life; the use of multiple SIM cards in mobile phones, for example, allows individuals to carefully manage their accessibility at different times and in different places (Green 2002; Green, Lachoee, and Wakeford 2001); instant messaging similarly offers new ways to manage presence and negotiate participation in different social groups (Grinter and Palen 2002; Consolvo et al. 2005). The issues of identity work in information practices and the assumptions behind technological designs are perhaps most strikingly illustrated by looking at non-Western cultures (Bell 2006b). As we suggested earlier in this chapter, the ideas of privacy inscribed in our technologies are derived largely from a Western context in which the individual human is the natural unit of social activity and analysis. In cultures where the family, household, or lineage group is more significant, however, the information-flow boundaries are radically different. These differences underscore the central observation that information technology does not simply encode social practice but instead is a site at which it is developed.

What we might be left wondering, in the face of these complexities, is whether the term "privacy" is actually useful at all. Perhaps it is so burdened with cultural specificities, legal interpretations, metaphoric resonances, and historical baggage that it obscures rather than illuminates.

Recent legislation in the United States has, in many states, mandated the lifetime tracking of paroled sex offenders with ankle-mounted GPS units in order to enforce parole conditions that prevent them from living or spending time near public parks, playgrounds, schools, and similar areas (Troshynski, Lee, and Dourish 2008; Shklovski et al. 2009). When these laws were initially passed, civil rights advocates challenged them in court on the basis that they constituted a curb on the right to privacy. Those legal challenges were unsuccessful; state appeals courts have ruled, in other words, that privacy is not a valid concern when considering these parolees' case. The question is, What concepts *are* usefully relevant, if privacy is not?

The parolees' experience is an illuminating one for many of the issues that we have described throughout this book. Navigating everyday space, organizing activity, and interacting with others are all transformed, to some degree, by the use of the technology, the representational frameworks it creates, and the forms of surveillance associated with them (Curry 2005). Most particularly, though, for the parolees and the parole officers whose job it is to monitor the parolees, many considerations arise that are deeply connected to those that we have discussed in this chapter.

Identity work and the social markers of membership of certain groups are most certainly at work here—the parolees respond to, and in some cases resist, their characterization as high-risk offenders (or in some instances accept the label and reflect on the ankle unit as a reminder of their past offenses), and the visible presence of the unit transforms their interactions with others by marking them out as offenders. Likewise, the way that information about their location can become a resource in managing their interactions with others—such as, for instance, the use of tracking data as a formal source of alibis and security against hassles from the police—is not dissimilar from the logics of risk and danger explored earlier.

Through this work, Dourish and his colleagues have suggested that the problems of privacy in location-based systems might be more productively framed in terms of the accountabilities to which people may find themselves held in their relations to others (Troshynski, Lee, and Dourish 2008). While parolees may have no right to privacy, it remains the case that they are accountable to each other, the police, the public, and the state for their presences and absences at different times; that their accountabilities to friends, family, and workmates may also be transformed; and that the accountabilities of those others toward the parolees are similarly changed. In other words, not only does the concept of accountability apply even when conventional social expectations of privacy have been ruled as inoperative but it is also a bidirectional relation that brings into focus the way that parolees and parole officers are mutually implicated in the technological regime (Shklovski et al. 2009). Parole officers similarly face a range of accountabilities to others, such as their superiors, the state corrections system, and the public at large.

The shift from a discourse of privacy to one of accountability perhaps gives us some purchase on those cases where privacy seems to "disappear" (e.g., Barkhuus et al. 2008). One of the common contemporary statements about privacy in digitally mediated settings—offered sometimes as a surprising observation, and sometimes as an operational rule of thumb—is that for particular people (often the young) in particular settings (often around new technologies), "privacy doesn't matter." For instance, when privacy questions are raised around specific prototype location-based technologies, such as those that share location information among friends, the rejoinder is usually that "people don't worry about privacy" in these cases, or within the constraints of a particular study. While these deployments genuinely do frequently fail to exhibit the kinds of concerns about surveillance and invasion that are stereotypically associated with privacy concerns, we find it useful to think of how, in fact, the disclosure of location information reflects accountability among peers.

Far from being able to choose not to disclose this information, a system might create a context in which one becomes accountable to others for the timely report of location. The issue is not whether people feel that their privacy is invaded but rather what kinds of social pressures they feel to share information. These social pressures are not purely coercive, of course. Nevertheless, they do reflect new and emerging accountabilities within social groups, as evidenced by the fact that a failure to provide an adequate, accurate, or timely account of one's location might be worthy of remark, if not censure. So the idea of accountability focuses our attention on what is achieved through location reports, in ways that the idea of privacy might not allow. This may prove a useful starting point for an account of the privacy problems of ubicomp, which rejects the notion of privacy as an analytic starting point.

Beyond Privacy

It is an article of faith in the ubicomp community—albeit one learned through some hard lessons—that in the design of an interactive system, usability cannot be an afterthought. A system cannot be made usable through the simple addition of a user interface, because usability is not limited to the user interface itself but a pervasive feature of system design. Privacy and security are similar; support for effective privacy protection cannot be "grafted onto" a system because it is a pervasive aspect of how that system is designed. In fact, as we have argued here, it is a pervasive aspect of how the system will be used and the context in which it is put to use, the values that it is used to support, the interpretations that others will make of its use, and so forth. Through a broad examination of related literature, we have been attempting to illustrate the inevitable social and cultural embeddedness of privacy and security questions and to draw out the consequences for how we talk about and design information systems. Privacy is clearly not a stable, universally understood phenomenon; rather, it is interpreted locally and through distinctly cultural lens.

Throughout this chapter, we have maintained that rational-actor economic models are inadequate as sole explanations of privacy and security practices, because they fail to capture other symbolic and social values of those practices. As Marshall Sahlins (1972) argues, social action is never purely utilitarian; it is culturally constituted. This is not simply an assertion that social factors are elements in the trade-off of costs and benefits. It is a claim that these are not individual decisions but instead collective actions, given form and meaning through how they produce and reproduce cultural and social values.

Second, we have suggested that information practice—collectively reproduced understandings of how information should be shared, managed, and withheld—may be more fruitful than traditional conceptions of privacy and security as ways to think about the broader context within which these issues are embedded. Turning away from privacy as an abstract goal and toward information practices as performative helps us to see how information is embedded in a wide range of forms of social action. From a design perspective, it calls into question the separation between configuration and action that characterizes most interactive systems for privacy and security management.

Third, we have shown that privacy practices are not merely ways in which information is managed but also ways in which social actions are achieved. Any adequate account of privacy behaviors, then, must be grounded in an understanding of the specific social and cultural context within which the activity is taking place.

Fourth, we have noted that the many different and dynamic social contexts within which people are embedded, as well as the very notion of practice, imply that information needs and uses are continually subject to change, revision, and reinterpretation. One implication of this is that models that require abstract specification (e.g., traditional access-control mechanisms and preferences) are inherently limited. Some have taken this as an argument in favor of machine learning, and related approaches can be applied to model context without relying on abstract models; but, again, the separation between representation and action renders these problematic as a means to enact information practice.

Fifth and finally, we have attempted to show that information practices cannot be separated from the concerns for risk, danger, trust, secrecy, identity, morality, and power that collectively give them meaning. Privacy is not a concept that can be separated from the collective practices through which it is achieved and made operable, nor from the other elements that are achieved through those same practices. Bruce Schneier (2000, 373) notes that "security mechanisms that aren't understood . . . by everyone don't work." We have tried to take this one step further, beginning to unpack the ways in which security and privacy, as information practices, are means by which people collectively understand the world.

The arguments that we have presented here are intended to further an exploration of privacy and security as social products rather than natural facts. As opposed to attempting to encode and replace individuals' or groups' information practices (as conventional privacy technologies do), we can seek to support and augment not just social practices but the means by which they evolve, too. In essence, our approach therefore seeks not to transform privacy into a technical property that can be automated but instead to support the human social and cultural practices through which the whole complex of phenomena—privacy, security, risk, danger, secrecy, trust, identity, morality, and power—are managed and sustained.

8 Domesticity and Its Discontents

Throughout our discussions, we have been considering ubicomp not simply as a new technology but also as a technological imaginary—something to think with, an idea that invites new sorts of speculation about what information technology might and could be. One of the central considerations presented by Weiser's initial account of the dawning age of ubicomp is to think about new arenas in which information systems might be deployed. One of these—present even in his original paper—is the home as a site of computation use.

This is well-traveled ground. Computer manufacturers have presented the home as a site of potential computer use for decades. In 1969, Honeywell famously offered its Kitchen Computer in the Neiman Marcus catalog (see figure 8.1). It was a floor-standing, low-powered minicomputer based on the sixteen-bit Honeywell 316 (one of the early ARPANET message-processing computers). Like other computers of the time, it lacked a screen or keyboard, presenting a user interface comprising only a few toggle switches and lights. In its Kitchen Computer configuration, the computer had its front panel extended to form a chopping board. It was advertised with the now-mind-boggling slogan, "If she can only cook as well as Honeywell can compute." The advertising image, showing the computer placed in the middle of a suburban kitchen, somewhat incongruous among the wooden cabinetry and floral-check wallpaper, and occupying more floor space than the refrigerator (whose door it blocks), does rather put the lie to any claim of practical domestic application (never mind the ten-thousand-dollar price tag). Yet the image stands more perhaps as a signal that the domestic sphere was already envisioned as territory ripe for computational colonization.

The Neiman Marcus appearance was among the earliest in a long series of advertising images of computers located in kitchens. The applications for the Kitchen Computer—storing and cataloging recipes, balancing the

If she can only cook as well as Honeywell can compute.

Her souffles are supreme, her meal planning a challenge? She's what the Honeywell people had in mind when they devised our Kitchen Computer. She'll learn to program it with a cross-reference to her favorite recipes by N-M's own Helen Corbitt. Then by simply pushing a few buttons obtain a complete menu organized around the entrée. And if she pales at reckoning her lunch tab, she can program it to balance the family checkbook. **84A** 10,600.00 complete with two week programming course **84B Fed with Corbitt data:** the original Helen Corbitt cookbook with over 1,000 recipes 5.00 (.75) **84C** Her Potluck, 375 of our famed Zodiac restaurant's best kept secret recipes 3.95 (.75) Epicure **84D Her tabard apron,** one-size, ours alone by Garden House in multi-pastel provincial cotton 28.00 (.90) Trophy Room

family accounts—are much the same as those on which the migration of computers into domestic space was predicated for over a decade after its introduction (although in time education and then games would come to dominate proposals for the home computer). To the extent that the kitchen is somehow emblematic of domestic life and presented as the hub of family activity, it stands for this broader argument. The ubicomp value proposition is somewhat different, in important ways. The home computer model suggests that the kinds of applications at which computers have excelled in business and scientific settings—information processing and cataloging, numerical processing—have some application in domestic settings too. In contrast, the much finer-grained ideas behind ubicomp—that we might benefit from a computationally augmented world rather than one with a computer in it—point to a deeper connection between information processing and domestic living (Tolmie et al. 2001; Crabtree et al. 2003). It is not merely the home that is to be colonized here by digital technologies but also domestic life itself. The relevance of this shift is that it is now domesticity—in all its social and cultural messiness and particularity—that comes into view.

Obviously, the Kitchen Computer was premised on a specific view of domestic life, but ubicomp by its nature depends on much more detailed speculations about the organization of domestic life. Ubicomp's accounts of domestic space and the opportunities for enhancements are by no means limited to technological commitments but also encompasses the social and physical constitution of the household; patterns of communication between family members; the rhythms and routines of home life; the interplay of leisure and work; the separations of activities; the temporal dynamics of the home over the course of days, weeks, months, and years; the boundaries between home and outside; and the movements of people, objects, and activities over those boundaries (Aipperspach, Hooker, and Woodruff 2008).

All this would perhaps be par for the course—ho-hum, another space to be filled by computational devices—were domestic spaces not subject to a quite different set of considerations than those that govern the offices, factory floors, and workplaces within which information technologies have conventionally been deployed. While the exigencies of global capitalism and the rigors of the marketplace mean that the dominant logic of workplaces is that of efficiency and cost-effectiveness, a different logic

Figure 8.1
Advertising image for the Honeywell Kitchen Computer. Reproduced courtesy of the Computer History Museum.

governs domestic life—one that is primarily emotional and moral rather than quantitative (although as Jean Baudrillard [1970/1998] famously observed, the logic of efficiency and performance is often invoked in the domain of leisure and consumption).

Critically, given the home as a major site of consumption, these normative accounts offer glimpses of domestic living as at once normalizing and aspirational. Jon Goss (1995, 191) provides an account of marketing as "an organized moral institution whose function is to promote social integration and control through the production of a system of coded values that are employed in the construction of . . . identities." From foodstuffs (and images of togetherness cemented by family mealtimes) to cleaning products (and images of love and protection through hygiene standards), the consumption of domestic goods and services is organized around moral narratives of appropriate family life. Information technologies are woven into these narratives, in both positive and negative terms (one must purchase Internet service in order not to place children at an educational disadvantage, but purchasing firewall software with "parental controls" is also necessary to protect children against both people and information that they might encounter online). So when ubicomp is figured as a technology for domestic space, it becomes part of this larger morality play.

In this chapter, we want to examine the interplay between technology and the home's social and moral organization, with a focus too on the fact that "the home" is a highly variable cultural object, in physical, social, economic, and emotional terms. We will approach this topic from three perspectives: the image of the so-called smart home and its connection to previous images of the "home of the future"; a decentered view of the home as a sociotechnical arena, concentrating on the boundaries of the movement of technologies into and out of domestic space; and considerations of safety and danger as they relate to information technology and nurturance.

Smart Home Deployments: Technologizing the House

Ghar. *결승점*. *Illam*. Council flat. Casa. Semidetached. *Haupt*. Town house. *Rumah*. Apartment. Sanctuary. Place of one's birth. RV. Refuge. *Danwei*. Purity. Modesty. A structure in which one lives. Simplicity. *Batch*. Family. Site of power. Memory. That there are so many words, metaphors, and imaginings for home should serve to remind us that homes exist within a wide range of physical, infrastructure, and legislative contexts and that they are also embedded within highly varied systems of meaning. Indeed, the world's 1.6 billion homes are as different from one another as the countries and cultures within which they were built. The structure and symbology of the home has been a classic topic of anthropological inquiry for decades—the symbolic organization of the Iroquois (Morgan 1851/2006) or Berber (Bourdieu 1970) house, or the colonial history of the bungalow (King 1984).

Materially, homes are hugely varied, and the challenges of designing for and into these many homes are immense. First, there are the practical considerations: size, density, scale, and history. The extraordinary variability in physical housing structures and types—from the vastness of U.S. homes with an average footprint of eighteen hundred to twenty-two hundred square feet to more modest homes elsewhere in the world—creates infrastructural, service-provisioning, networking, and device challenges. Increasing rates of urbanization worldwide, with its concomitant high-density dwellings, means that more than 50 percent of the world's housing stock is shared wall. Consider how that diversity in material form and structure meshes (or fails to mesh) with the design assumptions of domestic wireless-network access points. Second, homes are the sites of a range of social and cultural practices, dysfunctions, and aspirations, even within a single city (Miller 2009). There are a myriad of patterns of occupation, floor plans, household size, and composition. Single occupants, extended intergenerational families, and all points in between use the space within the home in complex, fluid, and culturally specific ways; front rooms, salons, and parlors might suggest one kind of hub within the home, while *anbangs* and antechambers point to another. Third, few homes operate in a vacuum or complete isolation; they are part of larger social, cultural, and sometimes physical institutions. Homes are connected up to other homes—some by proximity, others by shared infrastructures and resources, and still others by lines of affection and relationship. Lastly, and complicating the picture still further, the different kinds of metaphors and symbols of and for home mean that things we wrap around design, or that we imagine design might implicate—ideas about security, trust, the future, and even the relationship between public and private—are all flexible.

All this complexity seems to be at odds with the current, deceptively simple visions of the digital home. Not only is the home in these visions always singular, but it is nearly always unrealistically large, frequently freestanding, connected to the rest of the world only for the provisioning of services, and newly constructed—without legacy hardware, infrastructure, or quirks. It is also almost always occupied by a heterosexual nuclear family, which is remarkably accident-and trouble-free and perfectly happy to perform daily tasks and rituals in series or parallel, entirely without incident. The technology that makes this digital home smart is imagined to be seamlessly connected, robust, proactive, rational, and rationalizing, and ultimately delivering compelling experiences to its residents.

In the rush to sell the vision of the home as a site in which experiences could be enhanced through technology, the messiness of daily life was replaced with a vision of technological order. In some ways, this notion of the home as a site of enhanced experience is nothing new; nearly a century ago, Christine Frederick, Lillian Gilbreth, and Margarete Schütte-Lihotzky all concerned themselves with reimagining the home as a place of greater health, leisure time, and opportunities for self-improvement, and all drew on the principles of scientific rationalization that had developed in industrial settings. This vision would be achieved through the principles of domestic science and home economics, which brought scientific rationality to the home in the form of laborsaving devices. If labor in the home could be rationalized, these women argued, then we would free ourselves from all kinds of drudgery. Gilbreth was the wife and business partner of Frank Gilbreth, a pioneer in the scientific management movement spearheaded by Frederick Taylor. Famously, the Gilbreths used their own extensive family as a test bed for their theories. The fact that their autobiographical accounts of family life—*Cheaper by the Dozen* and *Belles on Their Toes*—continue to inspire reflections on the complexities of domestic living underscores the power of the vision of industrial order brought to domestic chaos. The temporal discipline of industrial capitalism identified by E. P. Thompson (1967) lives on through contemporary ubicomp projects and visions of the digital home (e.g., Harper 2003; Davidoff, Zimmerman, and Dey 2010).

Over the last decade, there have been a number of these smart home and digital home installations and experiments in the United Kingdom, Europe, the United States, and Asia. With one or two notable exceptions, these have been visions of domestic life that celebrated technology and its transformative power at the expense of home as a lived and living practice. We would argue that to design meaningful and meaning-making domestic technologies, one must begin with an awareness of cultural context, accrued social meanings, and everyday experiences. One must also begin with the home as it is currently built and experienced. In this approach, rather than imaging the home as merely another site of technology implementation, running parallel with the office, one starts with the lived home and augments it with digital technologies and infrastructures. Taking full and creative advantage of recent advances in technology and infrastructure—the proliferation and price stabilization of high-speed data connections, location-sensitive and location-aware devices, smaller-form factors, and underlying silicon advances (lower power requirements, or the continuing velocity of Moore's law)—will result, one hopes, in a larger portfolio of elaborations/interventions into the home as it becomes increasing digital.

Edge Structures: Enter the Shed

In his compendium on all things related to sheds in Australia, Mark Thomson (2002, 5) defines the shed as follows: "In Australia a shed can be anything from a dunny-sized [an outdoor toilet] construction to an aircraft hangar covering an acre or two. A shed might be defined as a building outside or away from the main domestic living space. . . . A garage is not a shed; it's simply a place in which to store a car." He goes on to argue that sheds transcend class and age: "Sheds are an integral part of Australian life. . . . In the shed, the rules are different. Here, chaos is allowed to reign, asserting its creative force in wayward contrast to the suburban order all around" (2–3). In the United Kingdom, writing in the new emergent genre of shed science or "shedist" literature, both Gareth Jones (2004) and Gordon Thorburn (2002) lay a similar claim to the importance of the shed in British culture, and its oppositional nature. Thorburn (ibid., 7) traces the shed's genealogy back to the Anglo-Saxon "scead" or shade, a place of partial darkness and obscurity, the realm of the hermit or wise man, and the storehouse for weapons and symbolic capital. This historical positioning of the shed does much to naturalize its role as a center of male power within the domestic sphere. Both Jones and Thorburn, and the British poet John Davies (2005), also bear photographic and narrative witness to the range of shed structures along with their continuing popularity in the United Kingdom—in 2004 alone, a record 1.5 million sheds were sold (Langley 2005).

Writing from our own homes, one in Long Beach, California, and the other in Forest Grove, Oregon, we are both currently without sheds. Yet as children of the former British Empire, growing up in Glasgow and in small and large Australian towns, we were both raised around shed culture: Dourish without a shed, but with a fascination for the small wooden garden shed next door with a smell of dirt that is not forgotten some thirty years later, and Bell with a range of sheds, including those in which her grandfather brewed beer, her father tinkered with steam engines, and her mother now keeps a lifetime's worth of academic documents and papers.

She also has the blueprints for the sheds that will someday rise behind her very American saltbox dwelling and her crumbling Australian stone cottage.

Clearly, as much as sheds function as sites of particular activities, they are also symbolic and cultural. They are indeed a cultural form, an imaginary realm within the larger domestic expanse as well as an explicit counterpoint to its more central elements. Michael Leunig (1990; see figure 8.2), a well-known Australian cartoonist, has several images of sheds in his work. In one, a woman sits on a sofa in the house reading *Mode* magazine and a man sits on a ratty chair in the shed reading *Mad* magazine; in another, two sheds stand on either side of a small winding road with a signpost that reads "SHEDVILLE Pop 2"—and a duck and a man stand disconsolate by their sheds. Significantly, these cartoons need no further explanation; that sheds are the moral domain of men and always accompany homes is implicitly understood, and they can thus be caricatured and quietly mocked. From this vantage point, one could suppose that sheds are a normalized part of the Australian suburban (and even urban) landscape(s). In fact, sheds figure in much Australian literature and popular culture, and similarly in the cultural production of the United Kingdom. In both traditions, sheds are often imagined as the sanctuaries of henpecked and downtrodden men, as a place for listening to Radio 4 or the Australian Broadcasting Corporation, potting and planting, or drinking and toiling over small projects away from the demands of family (Wyche, Taylor, and Kaye 2007). Sheds are portrayed as the centers of social life, places for shearing, preparing food on nearby barbeques, brewing beer, and making and playing music. Sheds are also constituted as a place of secrets and

Figure 8.2
Images of sheds in Michael Leunig's cartoons. Reproduced by permission of the artist.

things best left unspoken. In all of these imaginings, there are several constants. Sheds are physically separate from homes but make no sense without a home nearby, they are gendered male, and they seem to operate on altered schedules from that of the home.

Of course, the shed is an extremely flexible yet enduring category of domestic vernacular architecture. It can be as small as a lean-to or as grand as the collection of four-hundred square foot corrugated iron structures at the back of Bell's mother's garden in rural South Australia.[1] In Australia, some sheds can even be elaborate multiroom affairs. In the United Kingdom, where homes are half the size of those in Australia and where residential patterns tend to smaller land parcels, sheds are correspondingly less grand and might even be on land not adjacent to the home itself (e.g., allotment sheds, as discussed below). Still, within the British and Australian cultural traditions, sheds also function as real and imagined sites of technology consumption and storage within the larger domestic sphere. Tinkering in the shed, for instance, is one of the principle leitmotifs of innovation and invention in both Australia and the United Kingdom. Thomson (2002, 3), writing of Australian sheds, notes, "Our national knack for invention and innovation, for making do, lives on in the shed. The 'she'll be right' attitude may be denigrated as the light of Australian industry, but it thrives in the country's backyards." In some ways, one might also regard the shed as a real staging point for technologies coming into or out of the home; it is a place for not-yet-domesticated technologies, or for those that must forever remain feral and dangerous. Toward the end of *Blokes and Sheds*, Thomson draws an even more explicit parallel between sheds and technology. "Several of the stories," he writes of his collection, "hint at a new future for the shed-owning type of person: the computer. The virtual shed is with us now" (ibid., 255). Interestingly, in recent popular media coverage of the shed in Britain, there has been talk of wireless connectivity to the shed, with at least one cable company running a campaign to get so-called shedheads online with an information technology shed in north London (Wilson 2005).

The shed is scarcely the domestic space that comes to mind when we conjure up images of the technologized home of either the present or future. It is, as we have remarked, a space filled with technology, but it is not technologically enabled in quite the way that smart homes are meant

1. The English Heritage Trust currently lists more than fifty "sheds of special interest," and exhibits on sheds in Brighton as well as London's Victoria and Albert Museum have garnered considerable interest (Langley 2005; Davies 2005).

to be (Harper 2003). We find this absence telling. If the shed is a necessary part of the British and Australian domestic landscape, the masculine counterpoint to the perceived feminization of the home and the staging point or proving ground for technologies, where does it fit into visions of the homes of tomorrow? And if the shed is often an important boundary space for technologies—a site through which technologies move as they migrate both into and out of the home—then might it not provide us with a useful lens for thinking about how the smart home comes into being (and what might happen to it afterward)? The shed might provide a useful counterweight to the focus on the living room/kitchen or great room complex as sites of technology deployment (Mainwaring and Woodruff 2005). We want to lay claim to the shed as both metaphor and as site of technology consumption within the domestic sphere. In particular, we are interested in how sheds function at the edges of the domestic—their physical and potentially imagined life at the peripheries. The "edgefulness" of sheds might allow a different understanding of the home. Here the shed becomes a stand-in for a wide range of domestic but not home locations—the veranda, porch, patio, stoop, mudroom, nature strip, front yard, backyard, or deck. We are not suggesting a fundamental or wholesale rejection of the living room or other center points (Pullman-Jones 2005) but rather opening up the discussion of home to include these peripheries and edges.

The view from the periphery can be especially penetrating. Critical standpoint theory, including both feminist and Marxist articulations (see Haraway 1988; Harding 1991; Hartsock 1983, 1998), attributes epistemic privilege or authority based on a particular socially and culturally grounded perspective. In this theoretical construct, one's location as a social actor can mean a more or less acute understanding of larger social forces—that is, based on their experience as laborers, workers have an acute, and thus privileged, understanding of economics; women operating under patriarchy have a privileged understanding of gender in particular and society more generally. Standpoint theory is careful not to attribute this privilege or authority to any essential nature of laborers or women but rather to their lived experiences within larger systems of cultural production. Using (or inspired by) this theoretical lens, here we suggest that it is the marginal nature of the shed, and its existence on the peripheries of the domestic, that make it an ideal location from which to reexamine the home, its technological infrastructure, gendered divisions, and ultimately also its boundaries. We may gain more insight into the digital home by thinking about its edges as opposed to its centers. Our approach, then, is related to but distinct from the "extreme personas" employed by Tom Djajadiningrat,

Bill Gaver, and Joep Frens (2000), since marginality is crucial to our approach.

What Goes on in the Shed: The Shed as Lens

We would like to use the shed as a lens through which to scrutinize the conceptual frames within which technical conceptions of domesticity are situated. For us, the shed is both a part of the home and separated from it; the distinction between the shed and home thus draws attention to a series of otherwise-unstated assumptions about the nature of homes and domestic life. Exploring the role of the shed helps to unpack some of these assumptions. This is not a project about putting men back into the discussion of the home per se—that is ongoing elsewhere—nor is it about a celebration of paeans of British and Australian masculinity. Rather, we are interested in looking at the home as an imagined and real site of masculinity that is in turn not mapped onto established sites of public productivity (i.e., paid work and labor) or authority.

In this section, we examine a range of properties associated with the shed; in the next section, we return to the broader question of domesticity in light of these explorations.

Gender

Perhaps the shed's overriding feature as a cultural form is its gender orientation. Despite the rapid acceleration of female-controlled sheds in Britain (Langley 2005), the shed remains a thoroughly and quintessentially masculine space. It is a site of male habitation and practice, and many of the other elements that we will discuss with respect to sheds and domestic spaces—such as the association of sheds with tinkering, danger, and dirt—are strongly gendered too. The shed, in this view, cannot be taken as an object in itself but rather is enmeshed in a range of cultural oppositions that give it meaning. It is a primitive place with its deep, almost primal roots (Thorburn 2002); indeed, the very nature of its primitive form, from its construction to its contents, is one that helps to mark a range of domestic practice as female. In other words, sheds are sites for demonstrations and celebrations of certain forms of masculinity (that are themselves as stereotyped and culturally determined as the manifestations of femininity against which they are ranged). The pairing of "blokes and sheds" (Thomson 2002), then, is fundamental to the interpretation of sheds in relation to other domestic spaces. The shed is male space precisely because domestic space is largely read as female. It is a place where unfettered hypermasculinity rules the day, and to which masculinity may be exiled (see Refuge section below). Similarly, those objects and practices contained within sheds are equally male. In his remarkable short story *The Rough Shed,* the early Australian writer Henry Lawson offers this piercing articulation of the maleness of, in this case, a shearing shed. The sole protagonist and narrator, in reflecting on life in the shed, observes: "That's the way of it. If I went back to a woman's country again I wouldn't swear. I'd forget this as I would a nightmare. That's the way of it. There's something of the larrikin about us. We don't exist individually. Off the board, away from the shed (and each other) we are quiet—even gentle" (Lawson 1901, 275-276).[2] Here the shed is seen not only as masculine but also as collective and aggressive, and distinctly apart from women's domains.

Refuge

An important feature of the gendered nature of the shed—especially as it appears in fictional and literary accounts—is the shed as a place of male refuge from the house and family. The shed, in these accounts, is not merely a place for male work but more generally one for male isolation—a place for watching cricket, smoking a pipe, drinking, sleeping, quiet contemplation, and above all escape. In this view, the shed's space is not just male but more distinctly not female, it and becomes a space in which the feminine can be excluded.

Not surprisingly, in Britain, female interest in owning and occupying sheds engenders alarm (Langley 2005; Thorburn 2002), while in Australia the notion of the shed is being used now as part of a public health initiative targeting high suicide rates among men in rural areas (Leary 2005). A converted railway goods shed in Grenfell, rural New South Wales, was the first such "men's shed" where "guys meet other blokes, share a cuppa, have a yarn and when they unburden themselves they realize they're not the only one facing hardship. . . . The idea of the men's shed is a form of therapy, helping scores of men, 90% of whom have suffered some form of serious trauma in their lives, to be healed and come to a better place mentally" (ibid.). From this initial start, an entire men's shed movement has arisen, complete with websites, tax-exempt status, and factional disagreements.

If the shed is a place of danger in terms of the activities, objects, and substances that it contains, in terms of refuge it might also be seen as a

2. Of course, shearing sheds occupy a special space in Australian culture and history. Early unionization efforts were directed at shearing sheds and their shearers.

place of safety and comfort. It is a zone in which male pleasures will not be questioned, and in which a range of social norms and niceties might be relaxed or ignored altogether. It is a place where these activities do not seek female approbation and so do not risk female disapproval. Again, this is one way that the shed takes meaning from the series of cultural tensions it is embedded in and deeply reflects. The figure of the shed as refuge immediately raises two questions—Refuge from what? and Why?—and focuses our attention again on other domestic spaces in different ways.

The separation between the space of the shed and that of the house is perhaps even more marked in the British tradition, where sheds have historically been found on allotments—small plots of land rented out by local authorities to individuals for raising vegetables for domestic consumption. The typical allotment renter is someone who lives in a small house or apartment with no garden space of its own; allotments are centrally located on land controlled by the local authority and therefore physically separated from the home. The allotment is one of the natural habitats of the British shed, and its physical separation from the home only serves to emphasize its cultural and gender separation along with the notion of refuge that undergirds the movement from home to shed. According to William Langley (2005), however, "Allotments . . . have been disappearing at a terrifying rate as councils requisition the land for housing developments."

As concerns with sustainability and the environment have risen, the garden has slowly emerged as a site for technological interventions among a few researchers, some of whom recognize how gardens and gardening can also be sites for community activity (e.g., Goodman 2007; Pearce, Murphy, and Smith 2008). Articulating the relationship between home and garden, and building on all these specificities, remains challenging.

Migration of Technology

The location of the shed on the domestic sphere's edge also directs us to the ways in which the shed provides a space through which objects move on their way into and out of the home—a liminal space, if you will, maintained on the threshold of the domestic but where a different set of values and concerns hold sway. In their discussion of the notion of the "accidentally smart home," Keith Edwards and Rebecca Grinter (2001) discuss not just technologies within the home but also the manner and circumstances under which they come to be there. Sheds have a role here as spaces through which technologies move both into and, more commonly, out of domestic space.

The shed is often a place where junk accumulates—one of the characteristic images is of the accumulation of materials that have outlived their original purpose but are kept around in case a need someday arises. In this sense, then, the shed accumulates objects that are often derelict, discarded, or defective—objects with "spoiled identities" (Goffman 1963).

Since tinkering and "light engineering" are primary activities in the shed, defective technologies frequently move to the shed in order to be repaired. But since the aesthetic of the shed is that of the bricoleur too, the shed is also a place where old technologies are cobbled together, given new life, or pressed into marginal service. It thus functions as a channel by which materials move out of the house but not entirely out of the domestic sphere. By the same token, the shed may be also be a staging post in the movement of technologies into the home—a place where technologies and materials are, quite literally, domesticated and transformed to meet the needs of the home. The shed, then, figures in both the inward and outward migration of domestic technologies, and it is a critical staging post within their larger trajectories. Part of the domestic sphere yet separated from other domestic spaces, the shed safely contains technologies not quite ready for the home itself or past their usefulness.

Sheds as Chaotic Spaces

One image that frequently occurs in descriptions of sheds is of loosely ordered chaos. Sheds are messy spaces. The shed is a home for junk of all sorts, assorted odds and ends, and objects that are otherwise homeless. Sheds are inherently disordered. By the same token, when nothing has a place, then nothing can be out of place, and the shed is invariably a site of unexpected juxtapositions and combinations of objects and activities. In *Men and Sheds*, Thorburn (2002, 11) writes that "things accumulate in sheds, originally wanted then forgotten and much later resurrected as momentous mementoes." The image of the shed is one of a space piled up with building supplies, gardening supplies, tools, spare parts, scrap materials, household and industrial junk, leftover cans of paint, assorted screws and nuts and bolts, disassembled or half-assembled machinery, and almost anything else.

Again, the significance of this disorder can be seen in relation to the more ordered space of the home. Not only are homes, as public spaces, associated with the maintenance of particular kinds of order, but the home as a multiroom dwelling also allows for a segregation of activity between different kinds of spaces (rooms for eating, cooking, sleeping, socializing, etc.). Sheds are most often single rooms, for instance; there can be no segregation of activities or artifacts in such a space. In fact, the shed is resistant to such notions of order and accountability.

Sheds as Secret Spaces

Finally, another product of this separation between shed and domestic spaces, male and female spaces, and dangerous and safe spaces is the separation between a localized sacred and the profane. The shed is, to some extent, a space of secrets. Sheds are mysterious sites. To children, they are mysterious places from which they are largely excluded (so that, for example, to be granted rights to enter and use the shed becomes a rite of passage). Thorburn (2002, 7) describes the shed's contents as an assemblage of many things, some with magical significance, and his evocation of the hermit/oracle as the shed's ancestral inhabitant reinforces this lineage. In literature and fiction, the shed is often associated with secrets. Ada Doom of *Cold Comfort Farm* (Gibbons 1932) repeatedly and ominously intones that she "saw something nasty in the woodshed," something that had profoundly touched her life and that of her family; in the British film *Shaun of the Dead* (Wright 2004), the shed becomes the final resting place of Shaun's best friend and remaining zombie, where he can be imprisoned for his own good with a television to keep him company. Like the attic, the shed is a place where people, objects, and actions can be shut away from prying eyes and kept hidden from view. To paraphrase wildly, what happens in the shed, stays in the shed.

Home Sweet Home

According to annual crime statistics, Irvine, California, is the safest large city in the United States. All the same, danger remains a feature of the daily discourses in and around the city, as it does in most places. What is interesting to explore, though, is the ways in which dangers are formulated and projected. Like many conversations and imaginings of danger, the speakers—those assumed to be vulnerable, and those empowered to regulate—who monitor or remedy the situation have incredibly different perspectives. In chapter 7, we noted Day's research (2001) examining this issue by analyzing men's attitudes toward women's safety in public spaces in Irvine. What emerges from her work is not simply an interpretation of the potential dangers that exist in urban and domestic spaces but also an understanding of how these dangers are strategically deployed. In Day's study, the potential dangers to women in public space are a means by which women's movements in the urban environment can be controlled and curtailed. When certain places, activities, and times are marked as dangerous, then women's presence there—particularly on their own—is ruled off-limits. At the same time, the dangers to women that circulate as cultural values provide the opportunities for defining and exhibiting forms of masculinity—including examples that Day terms "chivalrous" masculinity (as marked by men accompanying women to dangerous places) or "bad-ass" masculinity (as marked by male presence in dangerous places).

Our concern is not with what danger is per se but instead with what danger *does*—danger as a social imaginary, cultural signifier, form of social regulation and control, and means by which cultural values are produced. Here danger operates not just as a lived reality but also a set of metaphoric and symbolic registers. In design contexts, danger frames a range of discussions of what technology might do, and the settings and relationships that require design attention and intervention, but rarely drive technology deployments. In our own work, we have been interested in exploring these questions in the context of current discussions of ubicomp. From our different perspectives—one writing as an anthropologist located in industrial research, and the other as an academic computer scientist—we find that the figuring of danger in contemporary technological discussions often obscures as much as it reveals.

In particular, as demonstrated by Day's example, when we talk about danger, we also invoke a series of cultural expectations and social groupings that we project onto the world. Take, for instance, any of the recent debates around such disparate topics as international and intranational terrorism, genetically modified foods, flu pandemics, renegade nation-states with nuclear weapons, closed-circuit television surveillance systems in public spaces, national identification cards, RFID-enabled passports, and MySpace, a social networking website where it is alleged that pedophiles regularly gather and one can see the wide-ranging set of practices that are routinely glossed as "dangerous." As these examples demonstrate, when we think about dangers—individually and collectively—our discussions are inevitably enmeshed in accounts of identity, individuality, community, gender, nationality, ethnicity, age, religion, and more, and those conversations take different forms in different cultural settings. Not surprisingly, these selfsame discourses around danger are even more visible when one is writing and thinking about technology in domestic and urban spaces.

During the second half of the twentieth century, computation moved from military to governmental to commercial settings; the office and factory floor were the primary sites for computerization over the last several decades. The early years of this century have been marked by a migration of computational technologies into new settings, not least of which is the domestic. Much research on ubicomp, then, begins with a picture of what domestic life might be like and by whom it is shared. Yet it runs the risk of falling prey to a variety of illusions about the nature of domesticity.

The splash page for the National Domestic Violence Hotline, a federally funded website in the United States that offers support to victims of domestic violence and abuse, includes the following warning: "Safety Alert. Computer use can be monitored and is impossible to completely clear. If you are afraid your internet and/or computer usage might be monitored, please use a safer computer, call your local hotline, and/or call the National Domestic Violence Hotline." Here the constellation of danger, technology, and the domestic is abundantly clear, albeit not entirely straightforward. According to U.S. government statistics, nearly four million women in the country experience a serious assault by a partner during an average twelve-month period, yet for as much attention as this issue routinely receives, violence, abuse, rape, and incest are not the experiences that comfortably populate our research imaginings of the domestic. Of course, for many occupants of domestic spaces in the United States and far beyond, these are the realities of daily life. In this framing of the domestic, homes are the places where women (and arguably children and older adults) are the most vulnerable; according to one set of statistics, women are most vulnerable to attack in their own homes between 6:00 p.m. on a Friday and 6:00 p.m. on a Sunday. Home is not sweet home; rather, it is a complicated and fraught landscape, negotiated with one eye to the nearest exit. In much of the writing about and designing for smart/digital homes, however, this set of realties is erased in favor of depicting occupants as a happy heterosexual nuclear family, with its safe relationships and sanitized patterns of occupation.

In this smart home, the pressing issues are of device controls, home automation, media circulation, and individual experiences of seamless connectivity. A casual survey of populist accounts of smart homes, from William Gibson to Hollywood, shows a remarkable preoccupation with self-cleaning systems and robotics (e.g., *The Jetsons* or *The Glass Bottom Boat*). A perusal of the research literature would suggest that the primary problem for ubicomp technology to solve in the home is the proliferation of media devices that clutter the living room, or perhaps something to help us remember our recipes (a target of computational design reaching back to Honeywell's Kitchen Computer in 1969 and earlier). Remarkably absent from these design proposals are, for example, technologies of domestic surveillance that might be used to detect the instances of teenage drug use, sexual predation, or domestic violence that statistics would indicate are more serious domestic problems than losing the remote. Clearly we are not proposing here that all smart homes should be designed to preclude illegal, immoral, or inappropriate behaviors but instead acknowledging that such things routinely transpire in homes, and might lead to different design choices and outcomes. A much different view of domestic technology might accompany an understanding of homes in terms of the dangers they potentially enclose, rather than those that they must be designed to exclude.

As noted in an earlier section, one way in which the gendered nature of the shed is worked out is in terms of the dangers associated with it. The shed typically houses a range of objects and substances that might maim, injure, or kill: "Danger lurks. Hidden in rough painted cupboards are forbidden chemicals. Rusting nails lie scattered around on the floor, ready to pierce unprotected feet. Screeching power tools send sparks flying. Risk and thrills are everywhere" (Thomson 2002, 2). The shed is the place where one might encounter rat poison, weed killer, acids, sharp blades, unprotected machinery, and live wires. The shed is also the place where these objects might be piled up precariously, dangled from hooks, kept in unlabeled boxes and bottles, and generally stored in flagrant disregard of tame, conventional safety procedures.

In the British lived experience, the shed is site of actual as well as perceived danger. The Royal Society for the Prevention of Accidents (2002) tracks household and leisure accidents and makes that information publicly available in annual reports on its website. In 2002, the last year for which data were available, more than 36,000 accidents culminating in visits to emergency rooms happened in garages, greenhouses, tool/potting/coal/shed/outbuildings, and the like. While this pales in comparison to the number of accidents in, for instance, the kitchen (261,949) or the bathroom/toilet (94,854), and is only a small percentage of the population (60 million), it still signals a source of realized dangers.

The association of danger with the shed depends on the relationship between the shed and other domestic spaces (Douglas 1966). The shed is the place to which those objects too dangerous to keep in the house are exiled. The purity of the house depends on the containment of dangers in the shed—which in turn seems to rely on rusted locks and aging padlocks. The shed is not only a container for dangerous objects but also a place for dangerous practices. Indeed, and in keeping with the hypermasculinity alluded to above, the shed is a place where dangerous practice is celebrated and ritually performed. Bottles of beer and rat poison sit side by side precisely because of the danger that is implied; knives, saws, and drills are scattered around, becoming playthings for children. The shed is not a place for safety goggles; it is a place to "get things done," and the fact that one might be continually in danger of losing a finger or an eye is part and parcel of the process.

Of course, the association of danger with male spaces is not unique to this sort of practice but is instead a common observation in anthropological work (Herdt 1994; Bell 1983). The ritual divisions between male and female spaces are frequently maintained through the invocation of dangers associated with female presence in male space. It is arguably this highly ritualized division that propelled reports of Sharon Thomas, a Welsh woman accused of running an international pornography ring, on to the front pages of newspapers in 2000—since, after all, she was keeping it in the garden shed (Housewife's Garden Shed Porn 2000; Boulware 2000). The outrage seems to be directed as much at her misappropriation of the shed as at any other illegality or immorality.

Where do risk and danger feature in ubicomp discourses about the home? To the extent that the home is conceptualized as a place of safety, comfort, and protection from the risks that populate the world outside, we often find ourselves blinded to the dangers that lurk within. Homes are the primary sites of domestic violence and sexual abuse, after all. So how do our technologies reflect those concerns? Consider, for example, technologies for tracking and domestic monitoring (e.g., Davidoff, Zimmerman, and Dey 2010). These technologies are designed to solve problems of coordination and communication, but at the same time their blindness to the home as a place of danger as well as comfort opens up new sorts of risk. In abusive households, say, domestic violence often takes the form of emotional manipulation and imbalanced power dynamics. Control over the movement of others—over the places they go, the company they keep, and the activities they undertake—is one of the primary ways in which that abusive power is exercised. What does it mean, then, to provide a technology in which the monitoring of the movement of others is a central function? What is important here is not to ask, What happens if this technology falls into the wrong hands?, but rather to question whether this kind of monitoring reflects our picture of normal family life, and what the implications are of placing this image of domestic life at the center of our design space. These questions prompt us from time to time to think about the ubicomp technologies that we don't see designed—the infidelity detector, the location-based spousal abuse recommender, or the teen-drug-abuse sensor network.

What we find useful about danger here is its strategic deployment to organize aspects of the social settings into which technology is projected. As a design imaginary in ubicomp, for instance, danger is caught up in a web of relationships to pastoralism, urbanism, domesticity, family life, and the disintegration of "traditional" values in modern life. At the same time, by situating design responses within these kinds of narratives, it provides a means to homogenize and stabilize the complex, contested, and ambivalent settings into which technology might be introduced. To us, *that* seems dangerous.

After the Shed?

Where does this exploration leave us? What might these elaborations of the secret lives of sheds, smart homes, and safety have to tell us about ubicomp?

Domestic environments have been a major focus of research attention for the ubicomp community, as evidenced by any number of papers exploring the potential of "smart space" technologies (Brummitt et al. 2000), advanced entertainment systems, group communication technologies (Hindus et al. 2001), augmented refrigerators and refrigerator magnets (Eardley et al. 2005), and so forth. Ubicomp is imagined here as a part of future homes. Ironically enough, the home of tomorrow is an old idea (Horrigan 1986), and like much science fiction it tends to owe more to our conceptions of the present than it does to likely futures. In using sheds as a vantage point from which to examine the ubiquitous home, what we hope to understand is what those images of future applications of ubicomp technologies to domestic spaces can tell us about the ways in which contemporary domesticity could be understood. What we find problematic about some of this work is that the image of the home as a physical space overwhelms or obscures the notion of the home as a social category. Any effective consideration of technologies for the home must be situated within understandings of the domestic as a form of cultural practice. In order to understand the domestic, we need to see how it is culturally produced and enacted in everyday practice, and how it emerges at a nexus of related concerns—such as around public and private, work life and family life, children and relatives, and responsibility and freedom. In other words, in understanding the home, what we turn our attention to is the way in which a domestic sphere can be distinguished from and articulated in relation to other forms of living and aspects of everyday life. Rather than assuming that the category of the domestic is pregiven and stable, we need to look at how it is actively produced.

For this reason, our methodological approach has been to focus on the edges and boundaries of the home—on how homes are maintained and how objects, people, and activities move across them. The edges are where the questions of the home are more contested and most visible, and places where we might inquire into such considerations as what constitutes home, family, and domesticity in the first place. As we have suggested, the shed occupies this border region, both spatially and conceptually. It is at the fringe of the domestic sphere. It partakes of the home but is also separate from it, and the segregation of activities, objects, and people between home and shed reflects a broader set of processes by which the boundaries of the home are maintained. What the shed provides, then, is a standpoint from which to reexamine the home.

What do we see when we look at the home from the shed? Many of the primary considerations are ones that we have prefigured in introducing and problematizing sheds in the first place, but let us recapitulate them here with a focus on home rather than shed.

How Things Move In and Out As others have noted, technologies do not simply appear in the home; they must be brought in, and then they must be domesticated. Similarly, the movement of technologies out of the home is one for which a range of trajectories can be articulated. Replaced or discarded pieces of tools, particularly in the case of expensive information technologies, typically find their way not to the dustbin but rather into other homes—such as those of friends, parents, and relatives. Hand-me-down technologies create gradients of sophistication that reproduce patterns of central and peripheral socioeconomic status. When thinking of how technologies move into and out of the home, what we see being produced are networks of homes and broader domestic spheres within which technologies are traded, and ways in which social ties are maintained by this circulation of technological goods (Malinowski 1922).

Gender and Technology One of the most surprising omissions from research on ubicomp in domestic space is the question of gender and technology (Wyche, Sengers, and Grinter 2006). The history of domestic technologies is a history of gender relations in the home (Cowan 1985). The introduction of vacuum cleaners, for example, turned carpet cleaning from a male task into a female one; similarly, advances in cooking and washing technologies, far from being "labor saving," served instead to transform expectations about culinary complexity and standards of cleanliness. In addition to thinking about how technologies move into and out of the home, then, one must also ask who brings them in, how they arrive, by what mechanisms they are domesticated, and what kinds of power displacements they achieve. For instance, Jennifer Rode, Eleanor Toye, and Alan Blackwell (2004) look at a range of programmable technologies in the home—not just computers, but VCRs, washing machines, and ovens, among others—to question who programs what and why. Perhaps unsurprisingly, what they find is a significant gender division in when people program, why they program, and what they program. When technologies are introduced into the home, they are introduced into gendered spaces, and they bring with them certain expectations about gender relationships. The family is both an inadequate unit of analysis and an inappropriate conceptual apparatus for understanding these questions.

Safety and Danger Homes frequently figure in ubicomp technologies as safe places. Whereas research into urban computing usually paints urban spaces as dangerous places to be either navigated quickly or in which you might want to be able to locate friends (and potential friends) among a host of hostile strangers, technologies of the home are ones that celebrate togetherness, warmth, and nurture. Technologies of the home are ones that promote sharing and togetherness (Strong and Gaver 1996), recording and revisiting special moments and celebrations (Truong, Huang, and Abowd 2004), or play and entertainment (Brummitt et al. 2000). At the same time, though, statistics show that homes are far from safe places. In 2003 in Australia, 24 percent of males and 47 percent of females indicated that the home was the site of their most recent assaults, with more than half knowing their attacker (Australian Bureau of Statistics 2003). In the case of sexual assaults, the figures are even more telling, with 86 percent of the victims of sexual assaults being women and 93 percent of the offenders being male. Forty percent of assaults took place in the victim's own or another home. In over half (58 percent) of the most recent incidents of female sexual assault, the victim knew the offender (ibid.). The statistics look remarkably similar from the United States and the United Kingdom. The home, it seems, can be a dangerous place.

Boundaries and Separations Rather than thinking about homes from the inside out, we find some value in thinking about them from the outside in. Whereas most inquiries into domestic ubicomp think about uniform access within the home, centered on such spaces as living rooms and kitchens, we want to think instead about the spaces where homes butt up against each other, where the boundaries must be maintained. Wireless networking technologies, with their implicit assumptions about the size and shape of residences, are one natural place where this concern with boundaries manifests itself; many technology adopters have stories to tell about interference between their own wireless devices and those of their neighbors, for example, and these problems manifest themselves differently in different countries with different forms of domestic infrastructure (Dourish and Bell 2007).

Tinkering and Technology Finally, despite the rhetoric of technology, tinkering, and end user programming of advanced technology in the home, it is notable that these activities are associated not with the home itself but instead with the marginal space of the shed. Again, Rode, Toye, and Blackwell's study (2004) points to a gender separation between different forms of programming and its different purposes. What we find intriguing here is how gender differences are reflected also in spatial segregations. Messing around with the technology—programming, tweaking, transforming, "opening it up," and "playing around"—are activities traditionally separated from the central spaces of the home, and indeed akin to work rather than to the domestic routine. Programming information systems realigns our relationships to it and the ways in which it can be embedded in domestic routines. The activities of the shed suggest that these transformations, for many technologies, are associated with an explicit movement of both artifacts and actions—a movement of the technology out of the home and an explicit process of reintroduction and reintegration later. While for many technologies programmability is a "feature" designed to encourage forms of domestic integration (e.g., Truong, Huang, and Abowd 2004; Humble et al. 2003), the view from the shed suggests that programming may well be incompatible with other forms of domestic practice, or at the very least that it implies certain patterns of ownership and control that need to be carefully thought through.

Our investigations here are motivated by the observation that in much research into domestic ubicomp, the term "domestic" is used largely as a taxonomic category for spaces, to distinguish them from those spaces that might otherwise be labeled as public, civic, workplace, commercial, or whatever. By contrast, we want to understand the cultural nature of the domestic itself—how domesticity is managed and produced. So we are interested in the social and cultural organization of domestic spaces and the nature of domesticity as a social imaginary (Taylor 2002).

One approach is to think about the center of the domestic—the concepts and spaces that are intrinsic to contemporary discourses about domesticity and home. The alternative approach that we have taken is to explore these questions from the outside and look at the ways in which the boundaries are policed. What we find is that the shed is a useful conceptual device for reframing technological discourse about the home. It serves this role not because of its centrality to the domestic sphere but precisely because of its marginal position. Environmental conditions mean that the shed takes on quite different physical forms in Australian and British domestic life, and yet it plays similar cultural roles. Australian and British sheds look quite different, house different artifacts, sit in different places, have different symbolic import, and so on. At the same time, though, they have much in common—not so much as artifacts in their own right but rather in terms of their positionalities within domestic ecosystems. They may speak in different accents, yet they have similar things to tell us about the homes in relation to which they are arranged.

The shed's position in the hinterlands of the domestic casts new light on the elements of domestic life that typically occupy design attention in ubicomp. In particular, it forces us to consider what sorts of margins surround the home, and in turn foregrounds a series of otherwise-unstated assumptions about the home and its borders—such as physical, group, practice, and conceptual borders. The home undoubtedly remains a fascinating site of investigation and opportunity for novel ubicomp technology, and much work remains to be done that takes the domestic as its central figure.

III

9 Reimagining Ubiquitous Computing: A Conclusion

Throughout this book, we have explored the ways in which ubiquitous computing has been imagined and instantiated over the last quarter century. Taking Weiser's defining "The Computer for the 21st Century" as our starting point, we have examined the silences, gaps, and erasures, and tried to find ways to bring new narratives, theories, and perspectives into dialogue with the more traditional formulations of a ubicomp agenda. Some will regard this as an explicit and hostile critique of ubicomp, or perhaps even its renunciation, but we do not see it this way. For both of us, ubicomp remains a fertile, productive site of inquiry. It is one of the few interdisciplinary hubs at which the intersections of new technologies and social practices can be theorized, built, and evaluated, then theorized all over again. This is possible in ubicomp through the deep entwining of social and technical in its most fundamental proposals, its close attention to emerging practice alongside technical innovation, and its embrace—always partial but nonetheless significant—of social, cultural, and humanistic inquiry alongside the technoscientific.

In part I, we outlined the methodological and epistemological commitments at work in our project. We explored two aspects of ubicomp. The first is the idea of ubicomp as a research and design project—one that animates a great deal of contemporary technological development. The second is the reality of ubicomp as a mundane element of everyday life for millions of people worldwide. Obviously, these are not entirely disconnected considerations; ubicomp developers are people too, and some of the seeds of tomorrow's everyday experiences lie in today's research laboratories. Still, the relationship between the two aspects of ubicomp is neither simple nor straightforward. For researchers and developers, ubicomp provides not only a set of goals but also a rhetoric, an agenda, and potentially a justification for projects, initiatives, and desires. For those of us who encounter computing technologies as somehow "ubiquitous," they may manifest themselves in terms of commitment and responsibility, amusement and diversion, or oppression and resistance, but always as objects to be picked up, adopted, adapted, and appropriated to personal and collective goals. We treat the two aspects here as somewhat independent concerns, however, not least because the relationship between them is a fruitful area of examination.

Having laid out this theoretical positioning, in part II we turned our attention to contemporary ubicomp practice. We did this by exploring a set of related yet distinct themes that arise both in the research literature and as aspects of everyday experience: infrastructure, mobility, privacy, and domesticity. These themes, by no means the only ones relevant to the ubicomp agenda, nonetheless did give us a clear and concrete set of sites at which we could interrogate the relationship between emergent technologies and daily life. Each theme provided us with an entry point and a particular set of phenomena that help to illuminate different aspects of the overall story. By weaving together a range of interdisciplinary approaches to these themes, we engaged in the sort of examination that we advocated toward the end of chapter 4—one that is grounded in empirical engagement with everyday life but that seeks within that an account of the patternings through which we construct coherence in the course of daily practice, with, through, and around technologies of all sorts, digital and otherwise. What they reveal is that when done right, ubicomp offers even more as a site of examination than simply an account of contemporary, cutting-edge technology development; it offers an examination of an ongoing working out of the nature of technologically mediated society. And it is this working out that excites us.

In this concluding chapter, we want to do two things: to articulate a clear framework for thinking about the kind of interdisciplinary engagements that this book exemplifies, and to build on that framework and point to some fruitful areas for ubicomp's next twenty years. In this chapter, then, we want to outline some of the characteristics of a new ubicomp agenda. This agenda should be read alongside and as a complement to such works as Adam Greenfield's *Everyware* (2006), Malcolm McCullough's *On Digital Ground* (2004), and Bruce Sterling's *Shaping Things* (2005). Our direction setting for a new ubicomp and its concomitant vision of a digital future is very much rooted in our unique positionality vis-à-vis this field. As an anthropologist and a computer scientist, in the university and in industry, both British Commonwealth expatriates living in the United States, we are aware that we have a particular set of intersections with the past, present, and future of ubicomp. Writing out of these subjectivities, we believe there are four distinct imperatives: a focus on everyday life and daily practice, an embrace of critical theory, an orientation to emergent technologies, and an awareness of new voices. But before we get to that we want to reflect a little further on the nature of the intellectual engagements that we believe will be necessary to engage with this set of imperatives.

Disciplinary and Interdisciplinary Engagements

The ubicomp research community has consistently avowed a commitment to interdisciplinary engagement, and appropriately so, since the very vision of ubicomp, as expressed by Weiser and growing out of Xerox PARC's research in the 1980s, was founded in a series of conversations and critiques implicating not only computer science but also psychology, sociology, anthropology, architecture, and other disciplinary positions. Consistent obstacles to the kind of interdisciplinary work that ubicomp posits, however, have included both the power relations implied by the dominance from engineering and computer science in ubicomp research (with all that implies for funding models, industry support, peer review practices, etc.) and a persistent confusion on the part of the dominant engineering community on quite what to *do* with critiques and analyses from other disciplines.

We tackle this problem in two ways. First, we articulate an alternative relationship between social theory and information technology design. Drawing on the arguments in chapter 4 about the analytic (rather than naively empirical) contributions of ethnographic research, we embrace Tara McPherson's notion of "theory objects." Here we use theory objects to suggest that technological artifacts operate not only as prototype commodities but also as embodiments of sociocultural theory. It seems to us that social science and ubicomp design practice may be brought together not purely in order to create usage models and personas but rather as a new, hybrid form of social and cultural investigation. Second, we illustrate how the positions that we have developed throughout this book and that run through all the topical themes in part II (specifically, a generative notion of culture, a processual account of everyday phenomena, and an examination of information technology as a site of cultural production), are implicated in current and future ubicomp design practice.

The central argument about the relationship between ubicomp practice and social science is that cultural phenomena are prior, not consequent, to design; so the goal is not to articulate the implications for design that follow from some understandings of the social but rather to examine and explore the implications for cultural embedding that are already inherent in design. Accordingly, while our position does not yield specific implications for the design of particular artifacts, it is aimed instead at producing a range of implications for ubicomp practitioners and practice.

As part of this examination of the implications of our work, it is appropriate to reflect on the project as a whole and, specifically, the form of interdisciplinarity that it represents. We are writing at a moment (if a rather extended one) when "interdisciplinarity," in name at least, is a byword in both academic and industrial research. The question of what interdisciplinarity might or should look like, up close, remains a question, and its value—beyond simple academic "marketing copy"—is often a similarly open question (Strathern 2004). Broadly, let us distinguish between two modes of interdisciplinary engagement: topical encounters and hybrid practices. By topical encounters, we mean those interdisciplinary projects that are united by a common but multifaceted phenomenon or topic of interest that is best understood through triangulation from the perspectives of different disciplines. Hybrid practices, by contrast, represents a form of engagement that creates new theoretical and conceptual frameworks bridging different disciplinary homes.

Within research domains such as ubicomp, there exist many topical encounters. There is a broad understanding that the phenomena of mobile, situated, and embedded digital technologies are sites for this kind of interdisciplinary engagements. Along the lines that we have discussed throughout this book, ubicomp technology is in practice as much a social and cultural phenomenon as it is a technological one; unsurprisingly, we find researchers from many different disciplines coming together under the umbrella of topics such as ubicomp, even if their methods, approaches, and analytic commitments do not always sit comfortably together. This perspective is often known as multidisciplinary rather than interdisciplinary research, but this distinction is not sufficiently shared or respected, in our experience, to give much useful purchase here.

These collaborations take place on many different scales—the domain, conference, workshop, project, or paper—and the scale shapes aspects of the engagement, but at its heart the recognition is that the topic goes beyond what a particular discipline can offer, so that multiple disciplines can be usefully brought to bear in concert with each other. It is this commitment to the idea that the topic is bigger than a single discipline can encompass that characterizes this style of interdisciplinary work.

Our own collaboration is certainly motivated by a belief that the phenomena of interest have escaped the bounds of any one disciplinary approach. But we approach this work from the perspective of an interdisciplinary hybrid practice rather than a topical encounter. For us, hybrid practice captures the sense that, as opposed to attempting to conduct work from our individual home disciplines alongside each other, we are conducting a new style of work that draws on each of our perspectives yet is reconfigured for the topic at hand—taking a sociotechnical perspective instead of studying the social and technical in parallel.

We should probably note that this was not necessarily a principled or carefully considered approach but rather one that we fell into easily, albeit motivated by aspects of our own previous research histories, institutions, and intuitions. As our project has progressed, however, it has become increasingly clear that it is important to explicitly engage in interdisciplinary practice rather than interdisciplinary projects, and to do so in ways that reflect the fluidity and hybridity of contemporary technological arrangements.

One particular form of hybrid practice has been exemplified by the approach we have taken throughout this book, which is neatly captured by McPherson's suggestion that technological and media artifacts can be thought of as theory objects. While technology developers think of new technologies, algorithms, designs, and systems as technological proposals—as solutions to or resolutions of past problems, and as potential products—we take them as embodiments of theories about social life and attempt to uncover, from those objects, what those theories might be. The relevance of this approach is that the theories represented by these objects are potentially those of many different parties: designers, certainly, but also users—individually and collectively—marketers, technologists and commentators, politicians and regulators, practitioners of various stripes, futurists and skeptics, and more. We take technological artifacts to be points at which the theory-making efforts of different groups overlap, intersect, and interact. It is not simply that social theories apply to technological artifacts but instead that they are already social theories, crystallized; similarly, it is not that their technological and social aspects can be understood separately but rather that they are simultaneously relevant.

This approach to technological artifacts as theory objects is, of course, intimately bound up with our concerns, as outlined in chapter 4, about the connection between social and cultural investigations and design practice. Specifically, we are concerned with the argument, implicit if not explicit in many discussions about the pitfalls of interdisciplinary investigation, that one primary measure of the strength of social or cultural investigation is the breadth of implications for design that result (Dourish 2006a). While we have both been involved in ethnographic work carried out for this explicit purpose, and continue to do so, we nonetheless feel that this is far from the only, or even the most significant, way for technological and social research practice to be combined. Just as from our perspective technological artifacts are not purely considered as "things you might want to use," from their investigation we can learn more than simply "what kinds of things people want to use." Instead, perhaps, we look to some of the questions that have preoccupied us throughout this book: Who do people want to be? What do they think they are doing? How do they think of themselves and others? Why do they do what they do? What does technology do for them? Why, when, and how are those things important? And what roles do and might technologies play in the social and cultural worlds in which they are embedded?

These investigations do not primarily supply ubicomp practitioners with system requirements, design guidelines, or road maps for future development. What they might provide instead are insights into the design process itself; a broader view of what digital technologies might do; an appreciation for the relevance of social, cultural, economic, historical, and political contexts as well as institutions for the fabric of everyday technological reality; a new set of conceptual resources to bring to bear within the design process; and a new set of questions to ask when thinking about technology and practice.

Technocultural Practices: Legibility, Literacy, and Legitimacy

We want to take a step back to consider a broader set of concerns within which the topics we have been discussing are embedded, which has proven useful in our own attempts to understand the relationship between the concepts we have outlined and design considerations. Again, for several reasons, our goal is not to provide specific guidelines for ubicomp design practice; such guidelines tend to be brittle and dependent on immediate technological opportunities rather than long-term concerns, for instance. Instead, what we have attempted to continually point to have been the ways that technological and cultural logics intersect. We approach this here in terms of three themes: legibility, literacy, and legitimacy. These themes provide orienting contexts.

Legibility

Digital technologies are framed in terms of information—information processing, seeking, management, retrieval, and sharing. Seeing information as a cultural category rather than a natural one, it may be more fruitful to think about how it is that people find spaces, settings, and activities informative. This redirects our attention from objects to processes, and to the relationship between forms of knowing, ways of being, and patterns of acting. In particular, it is useful to think in terms of the legibility of places, technologies, and actions—how it is that they can be read and understood as conveying specific sorts of messages.

The relevance of legibility lies primarily in the way that digital technologies can render the everyday world legible in new ways. Technology is a tool through which everyday practice can be read, processed, and understood. In this, we draw on many previous investigations that have taken legibility as a central concern. On an individual level, the legibility of urban space is the central topic of Lynch's *The Image of the City*, as we outlined in chapter 6. A form of collective legibility, though, is of more immediate relevance. The focus here is not on a personal experience; it concerns how social groups can share not only an experience but also a meaning for it. But going further, the ways we have of encountering space through practice are just that: ways we have. For Ito and Okabe's teens in Tokyo, the meaningfulness of mediated presence is a collective matter, not an individual one. Based on his work in southeast Asia, Scott (1998) discusses the history of the legibility of social life and attempts to control it. Two quite different forms of legibility arise from Scott's analysis.

One is what we call "panoptic legibility," the legibility of high modernism and central planning. In Scott's work, he associates this with modern statehood. For a state to control or manage (or exploit or appropriate) resources, it must first find a way to understand and compare those resources. Panoptic legibility is a centralized form, in which a standardized scheme can be applied across multiple settings and locales in order to measure and compare them. Standardized categories—whether of work or human action, land or natural resources, or whatever—can serve as the basis for understanding and allocation. Scott provides detailed examples, including agricultural or urban spaces laid out according to straight lines and right angles without reference to local topological features, and uniform single-crop (or single-strain) farming planned without reference to variable soil conditions or weather patterns. The primary characteristics of panoptic legibility are uniformity, abstraction, and dislocation; it is, almost by definition, a view from nowhere.

The alternative form of legibility explored by Scott is one grounded in indigenous practice, or what we term "local legibility." Rather than a view from without, this is the legibility of the view from within, the view "on the ground." Where panoptic legibility attempts to eliminate difference in order to achieve a coherent ordering of resources across different settings, local legibility focuses on the heterogeneous nature of everyday objects and actions, seeing them in terms of individual differences. Indeed, the forms of legibility at work here are heterogeneous. They exist at the intersection—or more appropriately, as the superposition—of many different systems. We mean this in two ways. The first is that there are many systems and infrastructures at work simultaneously, so Internet accessibility, mobile telephony, transportation systems, visual and physical access, and more all result in different forms of experience; consequently we must think of the ways in which they occur together. The second (reflecting the discussion of Massey's power geometries) is that experiences of systems and infrastructures are relative to the different constituencies, populations, and agencies at work. Most importantly, though, local legibility is the legibility of practice—it reflects how people work in, engage with, and make use of the world around them, rather than the abstracted view associated with panoptic legibility.

Scott uses the example of the contrast between Western single-crop agricultural management and indigenous African experiences of polycropping. To the Western eye, the practice of planting multiple crops in the same field or patch is disorganized and unscientific, lacking the precision that will allow for yield maximization. To the African farmer, on the other hand, polycrop farming is a practical way to ensure sufficient harvests in the face of poor soil and harsh weather, as well as providing for varied growth patterns that can help reduce erosion. An analysis of polycropping practices shows that the multiple crops are not planted at random but rather in careful relationship to each other, the local terrain and topology, soil conditions, historical patterns of crop success and failure, and so on. Like Western agricultural practice, polycropping draws on a complex store of knowledge and practice—but critically, it is one that is local, is grounded in the long-term, repetitive encounter with the environment, and operates on a different scale.

Informativeness and legibility are two sides of the same coin. The legibility of a system, setting, or activity is what allows us to find it informative and to see it as an instance of a category, as the kind of action that it is, as containing lessons, implications, or constraints. Legibility is a product of a social and cultural encounter with the world; in turn, it structures and shapes those encounters. Digital technologies that render the world legible in new ways—by creating new representational forms that account for action in the world (such as activity-recognition systems) or making the invisible visible (such as many location-based systems, social-networking systems, or data-mining applications)—help to shape this encounter and are themselves shaped by it.

The social origin of legibility is a critical issue for collaboration in mobile and ubiquitous environments. The examples that we have presented argue for a different view of information and information use than pervades conventional engineering discourse. They argue that the elements of the everyday world around which ubicomp applications seek to organize themselves—individuals, roles, groups, places, activities, times, contexts, and so forth—are not elements of the physical world to be uncovered and recognized but are instead elements of the social world. Their informativeness derives from the nature of social participation, and their nature and meaning are negotiated in, expressed through, and solely available to social practice. When we think of sensing technologies as devices that order the world, rather than devices that describe it, then alternative relationships between the social and technical are strikingly brought to light.

Literacy

The approach that we have adopted throughout the book emphasizes the processes by which our experience of the world is shaped and shared. That is, we take a practice-oriented view in which the ways of acting in different settings both reflect and sustain ways of understanding and organizing those settings. Applying this perspective to our conventional interpretation of information (and hence information technologies) has two consequences. One is that we should look toward the ways in which we must actively constitute the informativeness of the everyday world through our actions within it, and we explored this view through a series of examples considering how space might be found informative. A second implication, though, is worth raising here: the relevance of representational practices themselves. By representational practices, we mean both the practices by which certain kinds of representations are brought into existence, and the practices by which those representations are used, shared, and manipulated. Digital technologies are fundamentally representational—the basic stuff of computer programs is representations—and since ubicomp technologies focus especially on representations of the objects and activities of everyday life, the representational practice at work is consequential.

Walter Ong's classic account (1988) of the relationship between oral and literal cultures maintains that the different forms of representational practice associated with each result in quite different sorts of experiences of the world. The invention of written language allows for a form of static, reproducible, and transmissible experience of the past that is simply impossible to achieve in an oral culture. Looking from our own perspective, in which literacy is the basis of recorded knowledge, oral cultures seem simply to strive but fail to achieve the precision and durability of written knowledge. Yet Ong notes that the experience of the world in a preliterate culture is one in which no such durable, stable, and external record can exist; oral cultures are instead performative—ones in which, for example, poetic recitations are valued not for their accuracy but rather for their vibrancy and appropriate response to local conditions (indeed, to such a culture, accuracy would be viewed as a poor measure of aesthetic value, and not a part of the poet's art). At the same time, this performative nature of cultural knowledge is a source of reinvention and adaptation. In his study of Melanesian ritual practices, Fredrik Barth (1987, 31) ascribes certain aspects of the evolution of these rituals to the "repeated oscillations of cosmological lore between its private keeping and its public manifestations" associated with preliterate cultures.

Ong's focus on the performative aspect of oral culture clearly resonates with a processual account of information, but it also suggests a concern with similar aspects of written language. Written documents also have their performative aspects and, by extension, different kinds of representational forms, since they provide different sorts of orderings of objects and imply different kinds of understandings of the world. Jack Goody (1977) discusses different forms of knowing associated with basic literacy and later developments, such as lists and tables. In the absence of the list as a generalized form of knowledge, cataloging and ordering categories are not formalized as practices. Similarly, as the list emerges as a practical form, the practice of knowledge becomes the accumulation of lists, and then hierarchies, tables, and more. Studies of early book collections, such as the library of Elizabethan mathematician and magus John Dee (Sherman 1995), suggest that both forms of writing and even the physical forms of presentation themselves contribute to the practice of scholarship; if scholarship consists of amassing and assessing knowledge in the forms of books, then the forms of the books themselves and the capacities that they present—for marginalia, for endnotes, and so on—become aspects of the practice of scholarship and authentic knowledge.

In the spatial realm, of especial importance to ubicomp, maps are one of the most obvious intersections of practice, knowledge, and representation. The invention of maps gave rise to new ways of conceiving, cataloging, and moving through space, but maps carry with them commitments to forms of practice. Ed Hutchins (1995, 113) refers to navigational charts as "analog computers" for seafaring, noting that "not until the Mercator projection did a straight line have a computationally useful meaning." The cartographic projection with which we are most familiar, in other words, is designed in order to support specific kinds of navigational and computational practices. While a boon for Western navigation, the Mercator projection is a controversial one. In creating straight lines with navigational utility, the projection distorts the representations of Earth's surface area, exaggerating the size of countries that lie closer to the poles (largely first world countries and former colonial powers) while underrepresenting the landmass of those closer to the equator (often third world countries and sites of former colonial occupation). In this case, our appreciation of the vastness of the African continent is ruled as secondary to the opportunity to use geometric tools for navigation.

As a different form of cartography, consider the "occasion maps" (Garfinkel and Rawls 2002), or "mud maps," that one might draw when giving someone directions to a party or favorite coffee shop. What is represented here is not space but a journey, and we notate significant points along the way: landmarks and turns but not small bends in the road. Consistent representational schemes are forgone or transformed in support of the particular kinds of mutually understood practice within which the map will be put to use. This renders various ubicomp mapping and location-based projects moot, or at least it suggests that such endeavors need to privilege the narratives of directionality as much as the directions themselves.

For us, representational technologies are coupled with representational practices. Their accuracy or veracity can be defined only with respect to the specific practices by which they are employed and through which a relationship is established between the object and its representation. In the approach to information that we have been developing here, then, we similarly see the modern idea of information as a consequence of particular kinds of representational practices. Computer scientists and technologists read environments as informative according to a set of understandings they have of how the world might be represented; computational representations are tools of the trade, and learning to be a computer scientist involves learning to encounter the world as amenable to those sorts of representations, as a world of iteration and recursion (Lave and Wenger 1991). We make this point for two reasons. First, by denaturalizing computational representations and informational accounts of the everyday world, we want to further support a transition from information to informativeness, highlighting the role of mediating practices in informational accounts of ubicomp settings. Traditional informational accounts obscure the work that must be done in creating and maintaining a correspondence between the computational and noncomputational aspects of a setting (Smith 1996). Second, by emphasizing the processual aspects of information, we want to turn research attention toward alternative cultural experiences of settings in which technology might be embedded.

Legitimacy

Throughout this book, we have used examples of encounters with place and landscape to spotlight the variety of forms of "environmental knowing," suggesting that the account of information or knowledge incorporated in traditional technologies and technological representations is only one among many ways of understanding the relationship between people, space, and action. So, for instance, the cultural-historical landscape of the indigenous Australians and the moral landscape of the Western Apache sketched out in chapters 4 and 6, respectively, do not contain information in the manner we might normally propose but rather are inhabited in ways that render them informative. These alternative environmental epistemologies are products of habitation and purposeful action.

Nevertheless, there has been one significant consideration that we have not addressed, that these different epistemologies do not always sit comfortably side by side but instead are frequently in tension with each other. Implicit in any consideration of how to understand the informative nature of a space, then, is the question of the struggle for legitimacy of different forms of knowledge.

The context in which these struggles take place is the rise of technical rationality as the basis of both industrial practice and state governance. Management "by the numbers"—whether that is the management of production schedules, marketing campaigns, or state welfare—has become the dominant approach to understanding and acting within the natural world. Data analysis is the basis for understanding and responsiveness in this approach, and so information technologies of all sorts have played a critical enabling role (Yates 1993; Agar 2003). As scientific and computational accounts of the social and natural world are the basis of industrial and governmental practice, they inevitably come into conflict with the alternative epistemologies that they displace. Marilyn Strathern (2000) refers to one aspect of this broader phenomenon as "audit culture"—the movement of systems of measurement and quantification into all areas of state-regulated activity (such as education and the provision of local services), and the associated hegemony of league tables and numerical comparisons. Focusing on what he calls "virtualism," Daniel Miller (2003) draws attention to the movements of these models through the world; he notes that formalized models such as those that pervade audit culture tend to migrate toward centers of power, where they are often imposed on the world in ways that bring the world into alignment with the model, rather than the other way around. For instance, when the principles of neoclassical economics and macroeconomic modeling are adopted by organizations such as the International Monetary Fund and World Bank, they quickly move from being ways of describing the world to being ways of organizing it; projects such as the World Bank's structural adjustments program required governments to curtail those activities that would undermine or constrain the operation of the models. The legitimacy of representational and modeling approaches as embedded in systems of institutional power and authority are therefore highly consequential. It is critical to retain a focus on the contests of authority represented by the competition between different representational systems.

These issues are vividly demonstrated in disputes over First Peoples' land-right claims. In the United States, this has arisen as a problem of cataloging and assessing indigenous cultural resources. Richard Stoffle, David Halmo, and Diane Austin (1997) discuss this problem in relation to the Southern Paiute. The protection of cultural resources, when incorporated into Western scientific traditions, requires a means for calculating and comparing the cultural significances of different places, so that decisions can be made about priorities. Cultural resources are organized into "traditional cultural properties," which identify specific sites and objects of cultural significance for legal purposes. By contrast, Stoffle, Halmo, and Austin suggest that the Southern Paiute think not of specific properties, objects, or sites but rather of cultural "landscapes," which focus on the patterns of interdependency and proximity that link cultural resources rather than the properties intrinsic to one or another. Further, again, this holistic approach to the designation of cultural properties is one that is based around a human perspective instead of the "view from nowhere" of traditional cartography (Nagel 1986). So in addition to landmarks (which might fit within the traditional cultural property model), Stoffle, Halmo, and Austin point to the importance of holy landscapes, storyscapes, ecoscapes, and other ways of understanding the relationship between the land and cultural practice. Similar considerations have been documented in other groups, such as among the Navajo (Kelley and Francis 1993).

The Paiute example involves questions of one's claim on the land in the first place, but that is not our focus here. Nor is this simply a tale of incompatible ways of seeing the world. Rather, these instances are struggles for the legitimacy of different epistemologies (Eglash et al. 2004; Nader 1996). These different epistemologies are embedded within different systems of practice, and when the practices are in tension the legitimacy of forms of environmental knowing is called into question. Information technologies are those of representation, and as such they inscribe particular worldviews and inevitably obscure others. Information technology, tied as it is to our mental and cultural images of scientific representation and progress, is a tool not only for automation but also for legitimation.

A Reimagined Ubicomp

Legibility, literacy, and legitimacy, as we have articulated them here, are ways into thinking about as well as enacting a more fruitful set of interdisciplinary engagements around a future ubicomp practice. In what remains of this chapter, we want to return to the implicit project of the book itself—a reimagining of ubicomp.

We have used the phrase "mess and mythology" to characterize our investigations throughout this book. Our interest is not simply in each of these perspectives individually but rather in the consequences and conditions under which they occur together. When we talk of the mythology of ubicomp, our attention is directed to the way that the ubicomp project imagines the relationships between technology, people, and practice, both now and in the future. To the extent that ubicomp understands aspects of contemporary life to be not just problematic but specifically problematic in ways that are amenable to technological intervention and remediation, it both adopts and enacts a specific value system—one that animates and gives meaning to the world as we encounter it. When we speak of mess, our focus shifts toward the constant and ongoing complexity of the everyday world; the borderlands and hinterlands that hem in and surround the zones of technological development; the many different forms of life, work, and meaning layered in those spaces; and the ways that technologies and practices rely on a complex, barely understood, scarcely controllable patchwork of infrastructures that reach well beyond the "purely" technical to include, at the very least, educational institutions and practices, disciplinary conventions and boundaries, and market forces and the pressures of commercial finance. The inseparability of information technologies from the practices that bring them into being make them operational and give them local and particular meanings that yield a messy, tangled web. But it is when we bring these two—the mess and mythology—together that things get really interesting, because they struggle against each other. Messes challenge the myths and yet are shaped by them; mythologies bring order that messes resist.

When bringing this perspective to a study of ubicomp, we have been perhaps less concerned with the specific sites or topics of investigation as we have with the styles of investigation and the forms of scholarship that we have been trying to develop, exemplify, and advocate here. It is, for us, an approach entirely in keeping with the constitution of the ubicomp program as originally set out by Weiser. First, it attempts to take seriously what it might be to make a claim for a "ubiquitous" computing, not in the sense of subscribing to that as a plausible goal, but in understanding the ways that computing is both embedded within and constitutive of a world of shreds and patches, characterized by difference, disjunctions, and distinctions. The idea and practice of ubiquity—the work that needs to be done, by designers, users, legislators, and operators, to produce specific and limited forms of ubiquity—are themselves useful topics of interest, as are the areas of life, the peoples, the practices, and the worldviews excluded from this apparently expansive view. Taking ubicomp seriously, then, implies taking seriously the practices of ubiquity. Second, our approach draws on the broad intellectual vision that Weiser similarly sets out as a component of the ubicomp agenda from the outset. From the earliest accounts, ubicomp is no narrow topic located entirely, or even largely, within the boundaries of a traditional computer science. It expands those boundaries and reaches beyond them. As we have argued, we feel that this calls for integrative rather than parallel investigation from different disciplinary perspectives—another form of ubiquity perhaps, one we acknowledge as no less problematic.

Yet it is precisely these problems that make the area of ubicomp so productive. We suggested in the first chapter that what characterizes ubicomp settings is their messiness, and as we have proceeded through other topics, we have found this messiness to pervade methods, concepts, and practice. As we have observed, however, we are not saying that the messiness is an obstacle, nor are we advocating that it would be appropriate to tidy up. What happens if we allow that ubicomp will, by its nature as well as its connections to both people and technology, always remain messy? How would a do-it-yourself ubicomp be manifested? And what if we took these properties—instability, unreliability, and messiness—as important formulations rather than things that needed to be solved and addressed. We in fact find messiness inspiring, productive, generative, and engaging. Tidiness is static, rigid, fixed, and closed; messiness can be dynamic, adaptive, fluid, and open. Similarly, when we talk about the problems of ubiquity and the impossibilities of erasing difference, we do not see these as nails in ubicomp's coffin but instead as points of interest and exploration. Cleaning things up, then, has neither been our goal here nor has it become our goal going forward; embracing the messiness gives us so much more to think about.

In order to reimagine ubicomp we, perhaps ambitiously, want to suggest a new set of directions and underpinnings for the next quarter century. The ubicomp we are proposing is rooted in four things: a rich, nuanced embrace of the importance of daily life and everyday practice; a strong orientation to critical theory; a willingness to see technology in all its emergent messiness; and an ear for new voices. This is not a ubicomp that will discover lunchtime (O'Neill et al. 2006) or instrument yet another smart kitchen recipe finder (Ju et al. 2001; Tran, Calcaterra, and Mynatt 2005), and this is also not Weiser's ubicomp per se, with its relentless focus on embedded, distributed computational intelligence. One might argue, of course, that it returns to Weiser's original and novel impulses to marry computation with the world in which it could exist. This kind of relational framework is central to the ubicomp of interest to us. Ultimately we are concerned with the kind of messy, relentlessly unstable ubicomp that is produced when daily life and technology meet.

Indeed, one of the most fascinating elements of the ubicomp vision, and perhaps one of the reasons it is such a productive site for thinking about the nature of computational life, is that it is fundamentally relational. Previous approaches to thinking about computing have largely been defined in their own terms—in terms of computational power and functionality as well as capacities and capabilities. Ubicomp, on the other hand, is not simply about computing but also about the relationship between that computational capacity and the world in which it resides. What makes Weiser's vision unusual is that the story he tells is not only one of faster processors and algorithmic advances—a new *kind* of computing—but rather about the way in which computing will be woven into the fabric of everyday life—a new *place* for computing. The most significant thing to be said about ubicomp is where it is to be found—available beyond the desktop, carried with us, encountered as embedded in the built environment, and operating in the interstices between augmented artifacts. As ubicomp conjectures a world of ever-smaller artifacts, the analytic scope inversely increases to encompass all the other elements in the world, such as people and their bodies, conventions of physical space, and the practices that given them meaning and that are left out of many other areas of technology.

Clearly, as we have written throughout this book, ubicomp needs to be invested with a greater attention and have a more rigorous and theoretically constituted epistemological approach to making sense of people and their daily practices, as neither seem well understood or anticipated. The rise of new markets and market dynamics, especially in newly emerging economies and ones well off ubicomp's radar (i.e., India, Indonesia, etc.), suggests that there might be new questions to explore. Further, the appearance of new consumer forces (e.g., social media) and new cultural awarenesses and expectations of computing power and capability—privileging mobile devices, not fixed embedded intelligence, and deeply engaged with services and applications running on these devices in addition to the infrastructures undergirding the whole enterprise—means that the technologies of ubicomp will also look different moving forward. In what follows, we explore each of these imperatives in greater depth.

Daily Life and Everyday Practice

The expanding horizons of ubicomp's interest in daily life have begun to include a wider range of user populations, experiences, and sites of practices. The annual conference circuit and major publications increasingly reflect a broader set of practices, including topics around citizenship, civic participation, religion and spirituality, romance, and sex, or even a more explicit focus on work or leisure practices. Going forward, we hope to see our colleagues more fully examine these and many other facets of social and culture being.

Where does this interest in the "everyday" come from? Certainly, it was partly a consequence of the strong influence on HCI studies wielded by researchers who adopted an ethnomethodological orientation, with its interest in "mundane practices," while another influence involved those who work ethnographically, often highlighting the social and cultural complexity of otherwise apparently unremarkable activities. Partly it was the recognition that a diversification of forms of computational device might be accompanied by a diversification in the sites of their deployment. But one cannot help feeling, to return to Weiser's initial statement, that some of this was an explicit act of colonization—an attempt to seek out new markets for information technologies as the market for "office information systems" approaches saturation. In time, then, the identification of everyday life as a site of computational interest becomes something of a self-fulfilling prophecy. Importantly, though, it broadens the analytic focus of ubicomp as a domain by stressing the relevance of issues beyond system performance and usability. When everyday life is the object of attention, the question of how technology helps achieve human ends cannot be sidestepped.

In part II, we deliberately limited our attention to daily practices around privacy, domesticity, and mobility. We nonetheless embarked on a reading of these practices that could be a template for others. A profoundly different map of the world confronts the next generation of ubicomp research and researchers than that which Weiser was accommodating. The range and spread of digital devices and technologically mediated experiences are far beyond that of twenty-five years ago, and they include parts of the world and sectors of the population that were clearly well outside the frame in the early 1990s. Ubicomp must now contend with a diversity of domestic spaces and footprints, social organizational frames, social actors, work practices, and locations. As researchers, we will increasingly be confronted by questions about the nature of family, home, and work: Who is family? What is a home? Where do we work? These questions, which might historically have seemed relatively stable facts, become far more fluid and flexible when half of all Internet users are now located in mainland China not the United States, when the fastest-growing mobile phone markets are in Africa not Western Europe, and when social-networking sites rather than enterprise solutions dominate web traffic.

Moreover, it strikes us that studies of these complex patterns of social actors, contexts, and experiences are going to require a new set of conceptual tools and frameworks. It is safe to say that we will need to abandon the digital divide as one of the dominant organizing paradigms. The assertion of hyperconnected city dwellers and poor, frequently rural nonusers might not be nuanced enough to manage a world in which more than 50 percent of people will soon live in urban centers and where urban poverty is a real phenomenon. The presumed moral imperative around connectivity and access to the Internet as drivers to end poverty also probably needs further examination (Satchell and Dourish, 2009). How we account for, theorize about, and design toward those who are disengaged and differently engaged with new technologies is a fertile area for future inquiry.

Critical Theory

From its early incarnations, and in Weiser's own writing and talks, critical theory has been a part of the ubicomp rhetoric, albeit a small one. Indeed, many of the theoretical lenses we have evoked in this book were presaged by Weiser (1994)—feminist theory, postmodernism, and standpoint theory—but we would argue never developed to their fullest extent. Clearly we believe that critical theory is at the core of a reimagined ubicomp. And there is much that could and should be interrogated, such as sites of technology consumption, actors, motivations, and technology itself, not to mention the politics of consumption, bodies, design, and production.

We lowercase the term "critical theory" here—quite intentionally. Where Critical Theory tends to refer to one or another of a series of variously Marxist examinations of culture and ideology, those authors who bring a critical perspective to ubicomp analysis do so from a wide range of positions. In arguing for a central role for critical theory in a reimagined ubicomp, we want to create space for examinations of technology as a cultural force and of the conditions under which it takes shape. Studies of this sort are, of course, already well under way, and represented by the work of people such as Phoebe Sengers, Peter Wright and John McCarthy, Richard Coyne, and Phil Agre. In particular, Agre (1997) has articulated the idea of a critical technical practice, which points to the fact that the creation of new technologies and the exploration of them as cultural artifacts are not independent programs. Critical studies shape technological production; technological innovation is a response to cultural analysis; and most interestingly of all, information system design may itself provide a new mode of critical theorizing.

Throughout the book, we have relied on ethnographic material, both theory and methodologies, to shine a critical light on traditional ubicomp areas of concern—and this process helps us complicate the easy stories of mobility, domesticity, or privacy to suggest that these are by no means stable and fixed categories. Further, we have suggested that what makes them interesting is precisely this instability and fluidity; ubicomp will happen in the most intriguing ways and the messiest places. As a critical perspective argues, this messiness is a result of the meanings, conventions, and practices of ubicomp technologies being continually in flux. Ubicomp is a site of cultural production, and culture is dynamic and contested. The messiness is not a problem to be eliminated; it is an inherent property of the settings of technology use, and indeed, it's what makes things interesting.

There is, of course, more to ethnography than its ability to ground conversations in daily lived reality. It also has attendant to it a set of theoretical practices that have to underpin critical self-reflection. By this we mean the ability to talk about one's biography, location, and subjectivity, and the ways in which they might shape the identification of research problems, projects, and participants. While we would hate to see ubicomp practitioners and the field more broadly suffer through the "crisis of representation" that has beset anthropology, in particular, for the last twenty-plus years (e.g., Marcus and Fischer 1986), there is much to be learned from that process, and a small dose of critical reflection about subjectivity, positionality, and voice would go a long way. If we recognize that ubicomp design is doing cultural work—expressing and enacting values, shaping experience, and providing new sites and venues for interaction, negotiation, and imagination—then we can see the relevance of just how these technologies come to be. Ubicomp, like any technological domain, is continually oriented toward the future, but any thoroughgoing analysis must also be concerned with both its history (how we came to find ourselves where we are) and its contingency (how it might have been otherwise).

Emergent Technologies

Weiser placed his bets on sensor technology, embedded computational power, and the general dematerialization of devices, especially the desktop computer. We know now that while digital technologies did become ubiquitous, they followed a range of different arcs and trajectories, and any ubicomp agenda must contend with this multiplicity of future and stubborn current devices—televisions, smart electric meters, and automobiles, just to name a few. The computer is a general-purpose instrument, and so as technology researchers, we have a tendency to see these devices as computers, protocomputers, and quasicomputers, which can often blind us to the ways that these things are, as objects of human attention and interaction, quite different.

Furthermore, the ways in which the Internet has connected mobile devices to new streams of content has created new experiences and also new patterns of use and nonuse. It is thus increasingly clear that infrastructure, the mechanisms by which connectivity and content are delivered, are also implicated in ubicomp. More attention needs to be paid to how these operate and are delivered, billed, and regulated; to the ways they routinely fail, are thwarted, subverted, hacked, and repurposed; and, perhaps most importantly, to the ways in which they are imagined—to wit "cloud computing" or government-sponsored national broadband networks. Within any discipline, there is a natural tendency toward fragmentation and separation (Campbell 1969; Abbott 2001), so it is not surprising that topics such as contemporary interests in cloud computing are often hived off from domains such as ubicomp. But if we take ubicomp as a charge to think about computation ubiquitously—how it seeps into the various elements of everyday life, and how ubiquity is imagined as a goal—then we clearly need to resist the tendency to define cloud computing as a set of topics around parallel algorithms, process migration, server farms, and high-speed interconnects while ubicomp concerns itself with mesh networking, embedded devices, and sensor fusion. Cloud computing may not have been on the horizon when the initial ubicomp story was formulated, but it is central to any contemporary conception of what ubicomp might be.

The second consequence of this argument about emergence is that we need to return again to the lessons of the participatory design movement and its admonitions to question the boundaries of design. If design concerns the shaping of technologies to fit new needs and settings, then the design process is not finished when the last source file is compiled or when the product is packaged and shipped to a store. As Edwards and Grinter (2001) noted in their discussion of the digital home, our contemporary technological environments do not arrive all as a piece but rather are cobbled together bit by bit, each element pressed into service alongside the others—a wireless network hub here, a personal computer there, a digital video recorder, a camera, and so on. Technological environments are rarely designed in the classic sense; they are assembled through a process of bricolage.

Increasingly, too, design encompasses what we might call a digital afterlife. This includes the migration of secondhand technologies into aftermarkets and more distant sites (whether that means that a computer system has been gifted to grandparents, given to an after-school program, or pressed into service in another country), but it also encompasses the kinds of active repurposing associated with an emerging digital do-it-yourself movement along with the practices of circuit bending, hacking, and creative reuse. There is little published work in this space, but it is clearly an area of emerging research, which is, perhaps fittingly, appearing as workshops, installations, and hacks (e.g., Beuchley et al. 2009; Igoe 2007; Jungnickel 2007, 2008, 2009; Rosner 2010; Rosner and Bean 2009; Rosner and Ryokai 2008, 2009).

The broadening of focus that a concern with emergence suggests is both spatial and temporal—spatial in the sites of the practices that we find relevant for understanding ubicomp and its consequences, and temporal in the histories and afterlives of the kinds of technologies and technological practices that ubicomp considers.

Finding New Interlocutors and Voices

In the canonical research literature around ubicomp, such as the proceedings of the annual Ubicomp and Pervasive Computing conferences, there is a distinct overrepresentation of particular places, especially computer science departments, information schools, and industrial research laboratories in the United States and Britain. Increasingly, however, ubicomp voices can be found in organizational counterparts in Europe, Asia, and Latin America. As we have argued, the idea of what ubicomp might be—its prevailing mythology—can look quite different from these perspectives. Encouraging these new researchers and their divergent research projects is a critical part of a new ubicomp. Still, an opening up of ubicomp must also implicate colleagues in the social sciences, humanities, and arts and meaningfully engage with their disciplines and disciplinary concerns.

It also strikes us that the ubicomp of the next twenty-five years will have governments and government institutions as important stakeholders, interlocutors, and regulators. Globally, governments are increasingly concerned with the rapid proliferation of new digital technologies and infrastructures. In many instances, governments are the principle funders and drivers of such proliferations, which are underwritten by ideas of citizenship, productivity, innovation, and future proofing national markets. Many governments also regulate and constrain the spaces in which ubicomp will happen; governmental control or oversight of spectrum, content, and Internet provider addresses, for instance, could all have an impact on what ubicomp is and can become. What ubicomp's relationship might be to governments, beyond that of funding bodies, is then a space open for considerable debate.

One of our goals has been to lay the groundwork for encounters that might start to provide some answers in this area. In part, that has meant trying to set out and exemplify what an alternative ubicomp project might be, and to make clear the roles that other disciplines and stakeholders might play. In part, too, it has meant attempting to show to others outside the traditional ubicomp audience how research and development in the technical sphere is much more deeply entwined with, say, politics, governance, and policy than might otherwise be imagined.

One Last Thought

This exercise of spelling out a new ubicomp agenda could also be read as a larger charter. Throughout this book, we have tried to open up spaces where conversations about the future of technology can take place. It is our hope that by so doing, we can in some small way make room for the other voices, experiences, technologies, and theoretical directions that would make ubicomp even richer and more fertile territory. If Weiser envisioned ubicomp as a path toward "The Computer for the 21st Century," we are inclined to believe that it also lays the foundation for a diverse and similarly twenty-first-century scholarly investigation of computing.

References

Aarts, E., R. Harwig, and M. Schuurmans. 2001. Ambient Intelligence. In *The Invisible Future: The Seamless Integration of Technology into Everyday Life*, ed. P. J. Denning. New York: McGraw-Hill Companies.

Aarts, E., and S. Marzano. 2002. *The New Everyday: Visions of Ambient Intelligence.* Rotterdam: 010 Publishers.

Abbott, A. 2001. *Chaos of Disciplines.* Chicago: University of Chicago Press.

Abowd, G., C. G. Atkeson, A. F. Bobick, I. A. Essa, B. MacIntyre, E. Mynatt, and T. Starner. 2000. Living Laboratories: The Future Computing Environments Group at the Georgia Institute of Technology. In *CHI '00 Extended Abstracts on Human Factors in Computing Systems*, 215–216. New York: Association for Computing Machinery.

Abowd, G., and E. Mynatt. 2000. Charting Past, Present, and Future Research in Ubiquitous Computing. *ACM Transactions on Computer-Human Interaction* 7 (1): 29–58.

Abu-Lughod, A. 1986. *Veiled Sentiments: Honor and Poetry in a Bedouin Society.* Berkeley: University of California Press.

Ackerman, M. 2000. The Intellectual Challenge of CSCW: The Gap between Social Requirements and Technical Feasibility. *Human-Computer Interaction* 15 (2): 179–203.

Agar, J. 2003. *The Government Machine: A Revolutionary History of the Computer.* Cambridge, MA: MIT Press.

Agostini, A., G. De Michelis, M. Divitini, M. A. Grasso, and D. Snowdon. 2002. Design and Deployment of Community Systems: Reflections on the Campiello Experience. *Interacting with Computers* 14 (6): 691–714.

Agre, P. 1995. Conceptions of the User in Computer System Design. In *The Social and Interactional Dimensions of Human-Computer Interfaces*, ed. P. J. Thomas, 67–106. Cambridge: Cambridge University Press.

Agre, P. 1997. Toward a Critical Technical Practice: Lessons Learned in Trying to Reform AI. In *Bridging the Great Divide: Social Science, Technical Systems, and Cooperative Work*, ed. G. Bowker, L. Star, and B. Turner, 131–158 Mahwah, NJ: Lawrence Erlbaum.

Aipperspach, R., B. Hooker, and A. Woodruff. 2008. The Heterogeneous Home. In *Ubicomp '08: Proceedings of the Tenth International Conference on Ubiquitous Computing*, 222–231.

Anderson, K., and P. Dourish. 2005. Do You Know Where Your Mother [Trucker] Is? In *Proceedings of HCI International*.

Anderson, R. J. 1994. Representations and Requirements: The Value of Ethnography in System Design. *Human-Computer Interaction* 9 (2): 151–182.

Ang, P. H., and B. Nadarajan. 1995. *Censorship and Internet: A Singapore Perspective*. Available at <http://inet.nttam.com>.

Ang, P. H., Q. Zhou, and Y. Jiang. 2003. Lessons in Broadband Adoption from Singapore. *Journal of Interactive Advertising* 4 (1).

Aporta, C., and E. Higgs. 2005. Satellite Culture Global Positioning Systems, Inuit Wayfinding, and the Need for a New Account of Technology. *Current Anthropology* 46 (5): 729–753.

Appadurai, A. 1996. *Modernity at Large: Cultural Dimensions of Globalization*. Minneapolis: University of Minnesota Press.

Ark, W. S., and T. Selker. 1999. A Look at Human Interactions with Pervasive Computing. *IBM Systems Journal* 38 (4): 504–507.

Asokan, A. 2008. The Space between Mine and Ours: Exploring the Subtle Spaces between the Private and the Shared in India. In *Proceedings of Ethnographic Praxis in Industry Conference*, 175–191.

AT&T Laboratories Cambridge. 2002. The Active Badge System. Available at <http://www.cl.cam.ac.uk/research/dtg/attarchive/ab.html> (accessed October 17, 2008).

Australian Bureau of Statistics. 2003. *Bulletin no. 4509.0, Crime and Safety, Australia*. Canberra: Australian Government Printing Service.

Bardram, J. E., T. R. Hansen, M. Mogensen, and M. Søgaard. 2006. Experiences from Real-World Deployment of Context-Aware Technologies in a Hospital Environment. In *Proceedings of the International Conference of Ubiquitous Computing, Ubicomp 2006*, ed. P. Dourish and A. Friday, 369–386. Berlin, Germany: Springer.

Barkhuus, L., B. F. Brown, M. Bell, S. Sherwood, M. Hall, and M. Chalmers. 2008. From Awareness to Repartee: Sharing Location within Social Groups. In CHI '08: *Proceedings of the ACM Conference on Human Factors in Computing Systems, 497–506.* New York: Association for Computing Machinery.

Barkhuus, L., and P. Dourish. 2004, Everyday Encounters with Ubiquitous Computing in a Campus Environment. In *Proceedings of the International Conference on Ubiquitous Computing, Ubicomp 2004*, ed. N. Davies, E. Mynatt, and I. Siio, 232–249. Berlin: Springer.

Barley, S. 1986. Technology as an Occasion for Structuring: Evidence from Observations of CT Scanners and the Social Order of Radiology Departments. *Administrative Science Quarterly* 31: 78–108.

Barley, S., G. Meyer, and D. Cash. 1988. Cultures of Culture: Academics, Practitioners, and the Pragmatics of Normative Control. *Administrative Science Quarterly* 33 (1): 24–60.

Barth, F. 1987. *Cosmologies in the Making: A Generative Approach to Cultural Variation in Inner New Guinea*. Cambridge: Cambridge University Press.

Basso, K. 1988. Speaking with Names: Language and Landscape amongst the Western Apache. *Cultural Anthropology* 3 (2): 99–130.

Basso, K. 1996. *Wisdom Sits in Places: Landscape and Language among the Western Apache*. Albuquerque: University of New Mexico Press.

Baudrillard, J. 1970/1998. *The Consumer Society: Myths and Structures*. London: Sage.

Bazerman, C. 1988. *Shaping Written Knowledge: The Genre and Activity of the Experimental Article in Science*. Madison: University of Wisconsin Press.

Beck, U. 1992. *The Risk Society: Towards a New Modernity*. London: Sage.

Beck, U., A. Giddens, and S. Lash. 1994. *Reflexive Modernization: Politics, Tradition, and Aesthetics in the Modern Social Order*. Stanford, CA: Stanford University Press.

Becker, H. 1993. Theory: The Necessary Evil. In *Theory and Concepts in Qualitative Research: Perspectives from the Field*, ed. D. Flinders and G. Mills, 218–229. New York: Teachers College Press.

Beckman, C., S. Consolvo, and A. LaMarca. 2004. Some Assembly Required: Supporting End-User Sensor Installation in Domestic Ubiquitous Computing Environments. In *Proceedings of the International Conference on Ubiquitous Computing, Ubicomp 2004*, 107–124. Berlin: Springer.

Bedsworth, L., M. Lowenthal, and W. Kastenberg. 2004. Uncertainty and Regulation: The Rhetoric of Risk in the California Low-Level Radioactive Waste Debate. *Science, Technology, and Human Values* 29 (3): 406–427.

Bedwell, B., H. Schnädelbach, S. Benford, T. Rodden, and B. Koleva. 2009. In Support of City Exploration. In *CHI '09: Proceedings of the ACM Conference on Human Factors in Computing Systems*, 1171–1180. New York: Association for Computing Machinery.

Bell, D. 1983. *Daughters of the Dreaming*. Sydney: Allen and Unwin.

Bell, G. 2001. Looking across the Atlantic: Using Ethnographic Methods to Make Sense of Europe. *Intel Technical Journal* (Q3): 1–10.

Bell, G. 2005. The Age of the Thumb: A Cultural Reading of Mobile Technologies in Asia. In *Thumb Culture: Social Trends and Mobile Phone Use*, ed. P. Glotz, S. Bertschi, and C. Locke, 67–88. Bielefeld: Verlag.

Bell, G. 2006a. No More SMS from Jesus: Ubicomp, Religion, and Techno-spiritual Practice. In *Proceedings of the International Conference on Ubiquitous Computing, Ubicomp 2006*, ed. P. Dourish and A. Friday, 141–158. Berlin: Springer.

Bell, G. 2006b. *Satu Keluarga, Satu Komputer* [One Home, One Computer]: Cultural Accounts of ICTs in South and Southeast Asia. *Design Issues* 22 (2): 35–55.

Bell, G., M. Blythe and P. Sengers. 2005. Making by Making Strange: Defamiliarization and the Design of Domestic Technologies. *ACM Transactions on Computer-Human Interaction*, 12 (2): 149–173.

Bell, G., and P. Dourish. 2007. Yesterday's Tomorrows: Notes on Ubiquitous Computing's Vision. *Personal and Ubiquitous Computing* 11 (2): 133–143.

Bellotti, V., B. Begole, E. H. Chi, N. Ducheneaut, J. Fang, E. Isaacs, T. King, et al. 2008. Activity-Based Serendipitous Recommendations with the Magitti Mobile Leisure Guide. In *CHI '08: Proceeding of the ACM Conference on Human Factors in Computing Systems*, 1157–1166. New York: Association for Computing Machinery.

Bellotti, V., and S. Bly. 1996. Walking away from the Desktop Computer: Distributed Collaboration and Mobility in a Product Design Team. In *Proceedings of the ACM Conference on Computer-Supported Cooperative Work, CSCW'96*. New York: Association for Computing Machinery.

Benedict, R. 1946. *The Chrysanthemum and the Sword: Patterns of Japanese Culture*. Boston: Houghton Mifflin Company.

Benford, S., and L. Fahlen. 1993. A Spatial Model of Interaction in Virtual Environments. In *Proceedings of the European Conference on Computer-Supported Cooperative Work, ECSCW'93*. Dordrecht: Kluwer..

Benkler, Y. 2006. *The Wealth of Networks: How Social Production Transforms Markets and Freedom*. New Haven, CT: Yale University Press.

Bestor, T. 2004. *Tsukiji: The Fish Market at the Center of the World*. Berkeley: University of California Press.

Beuchley, L. E. Paulos, D. Rosner, and A. Williams. 2009. DIY for CHI: Methods, Communities, and Values of Reuse and Customization. In *Extended Abstracts of the ACM Conference on Human Factors in Computing Systems*, 4823–4826.

Beyer, H., and K. Holtzblatt. 1997. *Contextual Design: Defining Customer-Centered Systems*. San Francisco: Morgan Kaufman.

Bidwell, N., and D. Browning. 2006. Making There: Methods to Uncover Egocentric Experience in a Dialogic of Natural Places. In *Proceedings of the Australasian Conference on Computer-Human Interaction, OZCHI 2006*, 229–236.

Bittner, E. 1967. The Police on Skid Row: A Study of Peace Keeping. *American Sociological Review* 32 (5): 699–715.

Blank, J. 2001. *Mullahs on the Mainframe: Islam and Modernity amongst the Daubi Bohras*. Chicago: University of Chicago Press.

Blomberg, J., L. Suchman, and R. Trigg. 1997. Reflections on a Work-Oriented Design Project in Three Voices. In *Beyond the Great Divide: Socio-Technical Systems and Cooperative Work*, ed. G. Bowker, S. L. Star, L. Turner, and W. Gasser. 189–216. Mahwah, NJ: Lawrence Erlbaum.

Boas, F. 1940. *Race, Language, and Culture*. New York: Macmillan.

Boehner, K., R. DePaula, P. Dourish, and P. Sengers. 2005. Affect: From Information to Interaction. In *Proceedings of Fourth Decennial Conference on Critical Computing: Between Sense and Sensibility*. 59–68. New York: Association for Computing Machinery.

Boehner, K., J. Vertesi, P. Sengers, and P. Dourish. 2007. How HCI Interprets the Probes. In *CHI '07: Proceedings of the ACM Conference on Human Factors in Computing Systems*, 1077–1086. New York: Association for Computing Machinery.

Boellstorff, T. 2008. *Coming of Age in Second Life*. Princeton, NJ: Princeton University Press.

Bohannan, P., and M. Glazer. 1973. *High Points in Anthropology*. New York: McGraw-Hill.

Borriello, G., W. Brunette, M. Hall, C. Hartung, and C. Tangney. 2004. Remind about Tagged Objects Using Passive RFIDs. In *Proceedings of the International Conference on Ubiquitous Computing, Ubicomp 2004*, ed. N. Davies, E. Mynatt, and I. Siio, 36–53. Berlin: Springer.

Boulware, J. 2000. Tempest in a Garden Shed. Available at <http://archive.salon.com/sex/world/2000/07/07/shed/print.html>.

Bourdieu, P. 1970. The Berber House. In *Rules and Meanings: The Anthropology of Everyday Knowledge*, ed. M. Douglas, 98–110. Harmondsworth, UK: Penguin.

Bowker, G. 2006. *Memory Practices in the Sciences*. Cambridge, MA: MIT Press.

Bowker, G., and S. L. Star. 1999. *Sorting Things Out: Classification and Its Consequences.* Cambridge, MA: MIT Press.

boyd, d. 2007. The Cost of Lying. Available at <http://www.zephoria.org/thoughts/archives/2007/01/08/the_cost_of_lyi.html> (accessed October 18, 2008).

Brate, A. 2002. *Technomanifestos: Visions from the Information Revolutionaries.* New York: Texere.

Brenner, N. 1998. Between Fixity and Motion: Accumulation, Territorial Organization, and the Historical Geography of Spatial Scales. *Environment and Planning D: Society and Space* 16 (4): 459–481.

Brenner, N. 2001. The Limits to Scale? Methodological Reflections on Scalar Structure. *Progress in Human Geography* 25: 591–614.

Brewer, J., A. Bassoli, K. Martin, P. Dourish, and S. Mainwaring. 2007. Aesthetic Journeys: Rethinking Urban Computing. *IEEE Pervasive Computing Magazine* 6 (3): 39–45.

Brewer, J., S. Mainwaring, and P. Dourish. 2008. Aesthetic Journeys. In *Proceedings of the ACM Conference on Designing Interactive Systems, DIS 2008*, 333–341. New York: Association for Computing Machinery.

Brewer, J., and D. Nguyen. 2006. An Ethnography of OCTA Bus Travel. Working paper, University of California, Irvine.

Brown, B., and M. Chalmers. 2003. Tourism and Mobile Technology. In *Proceedings of the Eighth European Conference on Computer-Supported Cooperative Work,* eds P. Dourish, G. Fitzpatrick, and K. Schmidt. 335–354. Dordrecht: Kluwer..

Brown, B., and M. Perry. 2002. Of Maps and Guidebooks: Designing Geographical Technologies. In *Proceedings of the ACM Conference on Designing Interactive Systems, DIS 2002*, 246–254. New York: Association for Computing Machinery.

Brummitt, B., B. Meyers, J. Krumm, A. Kern, and S. Shafer. 2000. EasyLiving: Technologies for Intelligent Environments. In *Proceedings of the International Symposium on Handheld and Ubiquitous Computing*. London: Springer-Verlag.

Bruner, H. 1969. *Symbolic Interactionism: Perspective and Method.* Berkeley: University of California Press.

Buckland, M. K. 1991. Information as Thing. *Journal of the American Society for Information Science* 43: 351–360.

Bukatman, S. 1991. There's Always Tomorrowland: Disney and the Hypercinematic Experience. *October* 57 (Summer): 55–78.

Burrell, J., and G. K. Gay. 2002. E-graffiti: Evaluating Real-World Use of a Context-Aware System. *Interacting with Computers* 14 (4): 301–312.

Büscher, M., S. Gill, P. Mogensen, and D. Shapiro. 2001. Landscapes of Practice: Bricolage as a Method for Situated Design. *Computer-Supported Cooperative Work* 10 (1): 1–28.

Bush, V. 1945. As We May Think. *Atlantic Monthly* (July): 47–61.

Button, G. 2000. The Ethnographic Tradition and Design. *Design Studies* 21: 319–332.

Callon, M. 1986. Some Elements of a Sociology of Translation: Domestication of the Scallopes and the Fisherman of St Brieuc Bay. In *Power, Action, and Belief: A New Sociology of Knowledge?* ed. J. Law. London: Routledge and Kegan Paul.

Campbell, D. T. 1969. Ethnocentrism of Disciplines and the Fishscale Model of Omniscience. In *Interdisciplinary Relationships in the Social Sciences*, ed. M. Sherif and C. W. Sherif, 328–348. Chicago: Aldine.

Caspi, A., and P. Gorsky. 2006. Online Deception: Prevalence, Motivation, and Emotion. *Cyberpsychology and Behavior* 9 (1): 54–59.

Castells, M. 2000. *The Rise of the Network Society*. Oxford: Blackwell.

Chalmers, M., and A. Galani. 2004. Seamful Interweaving: Heterogeneity in the Theory and Design of Interactive Systems. In *Proceedings of the ACM Conference on Designing Interactive Systems, DIS 2004*, 243–252. New York: Association for Computing Machinery.

Chang, M., and E. Goodman. 2004. Fiasco: Game Interface for Location-Based Play. In *Proceedings of the ACM Conference on Designing Interactive Systems*, 329–332. New York: Association for Computing Machinery.

Chew, P. G. L., and A. Kramer-Dahl. 1999. *Reading Culture: Textual Practices in Singapore*. Singapore: Times Academic Press.

Choo, C. W. 1997. IT2000: Singapore's Vision of an Intelligent Island. In *Intelligent Environments: Spatial Aspects of the Information Revolution*, ed. P. Droege. Amsterdam: Elsevier Science.

Churchill, E., D. Snowdon, and A. Munro, eds. 2001. *Collaborative Virtual Environments*. Berlin: Springer.

Clifford, J. 1983. On Ethnographic Authority. *Representations* 2: 118–146.

Clifford, J., and G. Marcus. 1986. *Writing Culture: The Poetics and Politics of Ethnography*. Berkeley: University of California Press.

Consolvo, S., I. Smith, T. Matthews, A. LaMarca, J. Tabert, and P. Powledge. 2005. Location Disclosure to Social Relations: Why, When, and What People Want to Share. In CHI '05: *Proceedings of the ACM Conference on Human Factors in Computing Systems*, 81–90. New York: Association for Computing Machinery.

Cooper, G., and J. Bowers. 1995. Representing the User: Notes on the Disciplinary Rhetoric of HCI. In *The Social and Interactional Dimensions of Human-Computer Interfaces*, ed. P. Thomas, 48–66. Cambridge: Cambridge University Press.

Cowan, R. S. 1985. *More Work for Mother: The Ironies of Household Technology from the Open Hearth to the Microwave*. New York: Basic Books.

Coyne, R. 1999. *Technoromanticism: Digital Narrative, Holism, and the Romance of the Real*. Cambridge, MA: MIT Press.

Crabtree, A., T. Rodden, T. Hemmings, and S. Benford. 2003. Finding a Place for Ubicomp in the Home. In *Proceedings of the International Conference on Ubiquitous Computing, Ubicomp 2003*, ed. A. Dey, A. Schmidt, and L. Holmquist, 208–226. Berlin: Springer.

Crabtree, A., T. Rodden, P. Tolmie, and G. Button. 2009. Ethnography Considered Harmful. In CHI'09: *Proceedings of the ACM Conference on Human Factors in Computing Systems*, 879–888. New York: Association for Computing Machinery.

CRC (Censor Review Committee). 2003. Report of Censorship Review Committee, 2003. Singapore: Ministry of Information, Communication, and the Arts.

Crehan, A. C. 2000. Mabo-Style Claims to the Radio Spectrum. In *Selected Papers for the 2nd Australian Institute Conference on Computer Ethics*, 7: 24–28.

Cresswell, T. 2006. *On the Move: Mobility in the Modern Western World*. London: Routledge.

Cronon, W., ed. 1996. *Uncommon Ground: Rethinking the Human Place in Nature*. New York: W. W. Norton.

Curry, M. 2002. Discursive Displacement and the Seminal Ambiguity of Space and Place. In *The Handbook of New Media*, ed. L. Lievrouw and S. Livingstone, 502–517. London: Sage.

Curry, M. 2005. Toward a Geography of a World without Maps: Lessons from Ptolemy and Postal Codes. *Annals of the Association of American Geographers* 95 (3): 680–691.

Davidoff, S., J. Zimmerman, and A. Dey. 2010. How Routine Learners Can Support Family Coordination. In CHI 2010: *Proceedings of the ACM Conference on Human Factors in Computing Systems*, 2461–2470. New York: Association for Computing Machinery.

Davies, J. 2005. Shedman. Available at <http://www.shedman.net/index.html>.

Davis, F. 1959. The Cabdriver and His Fare: Facets of a Fleeting Relationship. *American Journal of Sociology* 65 (2): 158–165.

Day, K. 1995. Assault Prevention as Social Control: Women and Sexual Assault Prevention on Urban College Campuses. *Journal of Environmental Psychology* 15 (4): 261–281.

Day, K. 2001. Constructing Masculinity and Women's Fear in Public Space in Irvine, California. *Gender, Place, and Culture* 8 (2): 109–128.

de Certeau, M. 1984. *The Practice of Everyday Life*. Berkeley: University of California Press.

de Laet, M., and A. Mol. 2000. The Zimbabwe Bush Pump: Mechanics of a Fluid Technology. *Social Studies of Science* 30 (2): 225–263.

Djajadiningrat, J., W. Gaver, and J. Fres. 2000. Interaction Relabelling and Extreme Characters: Methods for Exploring Aesthetic Interactions. In *Proceedings of the ACM Conference on Designing Interactive Systems*, 66–71. New York: Association for Computing Machinery.

Douglas, M. 1966. *Purity and Danger: An Analysis of the Concepts of Pollution and Taboo*. London: Routledge.

Douglas, M., and A. Wildavsky. 1982. *Risk and Culture*. Berkeley: University of California Press.

Dourish, P. 2001. *Where the Action Is: The Foundations of Embodied Interaction*. Cambridge, MA: MIT Press.

Dourish, P. 2004. What We Talk about When We Talk about Context. *Personal and Ubiquitous Computing* 8 (1): 19–30.

Dourish, P. 2006a. Implications for Design. In CHI 2006: *Proceedings of the ACM Conference on Human Factors in Computing Systems*, 541–550. New York: Association for Computing Machinery.

Dourish, P. 2006b. Re-Space-ing Place: Place and Space Ten Years On. In *Proceedings of the ACM Conference On Computer-Supported Cooperative Work, CSCW 2006*, 299–308. New York: Association for Computing Machinery.

Dourish, P. 2007. Responsibilities and Implications: Further Thoughts on Ethnography and Design. In *Proceedings of the ACM Conference on Designing for the User Experience, DUX 2007*. New York: Association for Computing Machinery.

Dourish, P., and K. Anderson. 2006. Collective Information Practice: Exploring Privacy and Security as Social and Cultural Phenomena. *Human-Computer Interaction* 21 (3): 319–342.

Dourish, P., and G. Bell. 2007. The Infrastructure of Experience and the Experience of Infrastructure: Meaning and Structure in Everyday Encounters with Space. *Environment and Planning B: Planning and Design* 34: 414–430.

Dourish, P., and G. Bell. Forthcoming. "Resistance Is Futile": Reading Science Fiction and Ubiquitous Computing. *Personal and Ubiquitous Computing*.

Dourish, P., and G. Button. 1998. On Technomethodology: Foundational Relationships between Ethnomethodology and Systems Design. *Human-Computer Interaction* 13 (4): 395–432.

Dourish, P., R. Grinter, J. Delgado de la Flor, and M. Joseph. 2004. Security in the Wild: User Strategies for Managing Security as an Everyday, Practical Problem. *Personal and Ubiquitous Computing* 8 (6): 19–30.

Dourish, P., J. Lamping, and T. Rodden. 1999. Building Bridges: Customisation and Mutual Intelligibility in Shared Category Management. In *Proceedings of the ACM Conference on Supporting Group Work, GROUP'99*, 11–20. New York: Association for Computing Machinery.

Ducheneaut, N., and R. Moore. 2004. The Social Side of Gaming: A Study of Interaction Patterns in a Massively Multiplayer Online Game. In *Proceedings of the ACM Conference on Computer-Supported Cooperative Work CSCW 2004*. New York: Association for Computing Machinery.

Dumont, Louis. 1966/1980. *Homo Hierarchicus: The Caste System and Its Implications*. Rev. ed. Trans. M. Sainsbury, L. Dumont, and B. Gulati. Chicago: University of Chicago Press.

Durkheim, E. 1912/1965. *The Elementary Forms of the Religious Life*. English trans. J. Swaine. New York: Free Press.

Eardley, R., L. Swan, A. Sellen, A. Taylor, S. Hodges, K. Wood, and L. Williams. 2005. Augmented Refrigerator Magnets. Poster presented at Ubicomp 2005, Tokyo.

Edwards, P. 1996. *The Closed World: Computers and the Politics of Discourse in Cold War America*. Cambridge, MA: MIT Press.

Edwards, W. K., and R. E. Grinter. 2001. At Home with Ubiquitous Computing: Seven Challenges. In *Proceedings of the International Conference on Ubiquitous Computing, Ubicomp 2001*, ed. G. Abowd, B. Brummit, and S. Shafer, 256–272. Berlin: Springer.

Eglash, R., J. Croussant, G. Di Chiro, and R. Fouche, eds. 2004. *Appropriating Technology: Vernacular Science and Social Power*. Minneapolis: University of Minnesota Press.

El Guindi, F. 1999. *Veil: Modesty, Privacy, and Resistance*. New York: Berg.

Ellwood-Clayton, B. 2003. Virtual Strangers: Young Love and Texting in the Filipino Archipelago of Cyberspace. In *Mobile Democracy: Essays on Society, Self, and Politics*, ed. K. Nyiri. Vienna: Passager Verlag.

Ellwood-Clayton, B. 2005. Desire and Loathing in the Cyber Philipines. In *The Inside Text: Social Perspectives on SMS in the Mobile Age*, ed. R. Harper. London: Springer-Verlag.

Erickson, B. 1981. Secret Societies and Social Structure. *Social Forces* 60 (1): 188–210.

Euler, E., Jolly, S., and Curtis, H. 2001. The Failures of the Mars Climate Orbiter and Mars Polar Lander: A Perspective from the People Involved. In *Proceedings of Guidance and Control 2001, American Astronautical Society*.

Facer, K., and R. Furlong. 2001. Beyond the Myth of the "Cyberkid": Young People at the Margins of the Information Revolution. *Journal of Youth Studies* 4 (4): 451–469.

Facer, K., R. Sutherland, R. Furlong, and J. Furlong. 2003. *Screenplay: Children and Computing in the Home*. New York: Routledge.

Fine, G., and L. Holyfield. 1996. Secrecy, Trust, and Dangerous Leisure: Generating Group Cohesion in Voluntary Organizations. *Social Psychology Quarterly* 59 (1): 22–38.

Flintham, M., R. Anastasi, S. Benford, T. Hemmings, A. Crabtree, C. Greenhalgh, T. Rodden, N. Tandavantitj, M. Adams, and J. Row-Farr. 2003. Where On-Line Meets On-the-Streets: Experiences with Mobile Mixed Reality Games. In *CHI '03: Proceedings of the ACM Conference on Human Factors in Computing Systems*. New York: Association for Computing Machinery.

Floerkemeier, C., R. Schneider, and M. Langheinrich. 2004. Scanning with a Purpose: Supporting the Fair Information Principles in RFID Protocols. In *Proceedings of the Second International Symposium on Ubiquitous Computing Systems, UCS 2004*, 214–231.

Flyvbjerg, B. 2001. *Making Social Science Matter: Why Social Science Fails and How It Can Succeed Again*. Cambridge: Cambridge University Press.

Forsythe, D. 1989. It's Just a Matter of Common Sense: Ethnography as Invisible Work. *Computer-Supported Cooperative Work* 8 (1–2): 127–145.

Fosket, J. 2004. Constructing "High-Risk Women": The Development and Standardization of a Breast Cancer Risk Assessment Tool. *Science, Technology, and Human Values* 29 (3): 291–313.

Foth, M., ed. 2009. *Handbook of Research on Urban Informatics: The Practice and Promise of the Real-Time City*. Hershey, PA: Information Science Reference, IGI Global.

Foucault, M. 1975. *Discipline and Punish: The Birth of the Prison* [Surveiller et punir: Naissance de la Prison]. Trans. A. Sheridan. Paris: Gallimard.

Frost, P. J., and R. E. Stablein. 1992. *Doing Exemplary Research*. Newbury Park, CA: Sage.

Ganter, R., J. Martinez, and G. Lee. 2006. *Mixed Relations: Asian-Aboriginal Contact in North Australia*. Perth: University of Western Australia Press.

Garfinkel, H. 1967. *Studies in Ethnomethodology*. Cambridge, UK: Polity.

Garfinkel, H., and A. Rawls. 2002. *Ethnomethodology's Program: Working out Durkheim's Aphorism*. Lanham, MD: Rowman and Littlefield.

Garlan, D., D. Siewiorek, A. Smailagic, and P. Steenkiste. 2002. Project Aura: Toward Distraction-Free Pervasive Computing. *IEEE Pervasive Computing Magazine* (April–June): 22–31.

Garrioch, D. 2003. Sounds of the City: The Soundscape of Early Modern European Towns. *Urban History* 30: 5–25.

Gaver, W., T. Dunne, and E. Pacenti. 1999. Cultural Probes. *Interaction* 6 (1): 21–29.

Geertz, C. 1973. *The Interpretation of Cultures*. New York: Basic Books.

Geertz, C. 1988. *Works and Lives: The Anthropologist as Author*. Palo Alto, CA: Stanford University Press.

Geertz, C. 2000. *Available Light: Anthropological Reflections on Philosophical Topics*. Princeton, NJ: Princeton University Press.

Geiryn, T. 1983. Boundary Work and the Demarcation of Science from Non-Science: Strains and Interests in Professional Ideologies of Scientists. *American Sociological Review* 48 (4): 781–795.

Gershenfeld, N., R. Krikorian, and D. Cohen. 2004. The Internet of Things. *Scientific American* 291 (4): 76–81.

Gibbons, S. 1932. *Cold Comfort Farm*. London: Longman.

Gibson, W. 1999. Interview on NPR's *Talk of the Nation*, November 30.

Goffman, E. 1963. *Stigma: Notes on the Management of Spoiled Identity*. Englewood Cliffs, NJ: Prentice-Hall.

Goodman, E. 2007. It's Not Easy Seeing Green: Imaging Urban Parks within Neighborhoods. In *Proceedings of the CHI 2007 Workshop on Imaging the City*.

Goodwin, C. 1994. Professional Vision. *American Anthropologist* 96: 606–633.

Goodwin, C. 1995. Seeing in Depth. *Social Studies of Science* 25: 237–274.

Goodwin, C., and M. Goodwin. 1996. Seeing as a Situated Activity: Formulating Planes. In *Cognition and Communication at Work*, ed. Y. Engeström and D. Middleton, 61–95. Cambridge: Cambridge University Press.

Goody, J. 1977. *The Domestication of the Savage Mind*. Cambridge: Cambridge University Press.

Goss, J. 1995. "We Know Who You Are and We Know Where You Live": The Instrumental Rationality of Geodemographic Systems. *Economic Geography* 7 (2): 171–198.

Graham, S., and S. Marvin. 2001. *Splintering Urbanism: Networked Infrastructures, Technological Mobilities, and the Urban Condition*. London: Routledge.

Grannis, R. 1998. The Importance of Trivial Streets: Residential Streets and Residential Segregation. *American Journal of Sociology* 103 (6): 1530–1564.

Green, N. 2002. On the Move: Technology, Mobility, and the Mediation of Time and Space. *Information Society* 18: 281–292.

Green, N., H. Lachoee, and N. Wakeford. 2001. Rethinking Queer Communications: Mobile Phones and Beyond. Paper presented at the Sexualities, Media, and Technology Conference, University of Surry, June 21–22.

Greenfield, A. 2006. *Everyware: The Dawning Age of Ubiquitous Computing*. New York: New Riders.

Grinter, R., and L. Palen. 2002. Instant Messaging in Teen Life. In *Proceedings of the ACM Conference on Computer-Supported Cooperative Work, CSCW 2002*, 21–30. New York: Association for Computing Machinery.

Griswold, W., P. Shanahan, S. Brown, R. Boyer, M. Ratto, B. Shapiro, and T. Truong. 2004. ActiveCampus: Experiments in Community-Oriented Ubiqutious Computing. *IEEE Computer* 37 (10): 73–81.

Grudin, J. 1990. The Computer Reaches Out: The Historical Continuity of Interface Design. In *CHI '90: Proceedings of the ACM Conference on Human Factors in Computing Systems*, 261–268. New York: Association for Computing Machinery.

Gruteser, M., and D. Grunwald. 2003. Anonymous Usage of Location-Based Services through Spatial and Temporal Cloaking. In *Proceedings of the ACM/USENIX International Conference on Mobile Systems, Applications, and Services, Mobisys 2003*.

Gupta, A., and J. Ferguson. 1997. *Anthropological Locations: Boundaries and Grounds of a Field Science*. Berkeley: University of California Press.

Hancock, J. T. 2007. Digital Deception: When, Where, and How People Lie Online. In *Oxford Handbook of Internet Psychology*, ed. K. McKenna, T. Postmes, U. Reips, and A. N. Joinson, 287–301. Oxford: Oxford University Press.

Hancock, J. T., L. Curry, S. Goorha, and M. Woodworth. 2008. On Lying and Being Lied To: A Linguistic Analysis of Deception in Computer-Mediated Communication. *Discourse Processes*, 45: 1–23.

Hancock, J. T., C. Toma, and N. Ellison. 2007. The Truth about Lying in Online Dating Profiles. In *CHI '07: Proceedings of the ACM Conference on Human Factors in Computing Systems*, 449–452. New York: Association for Computing Machinery.

Haraway, D. 1988. Situated Knowledges: The Science Question in Feminism and the Privilege of Partial Perspective. *Feminist Studies* 14 (Fall): 575–599.

Harding, S. 1991. *Whose Science, Whose Knowledge? Thinking from Women's Lives*. Ithaca, NY: Cornell University Press.

Harper, R. 1992. Looking at Ourselves: An Examination of the Social Organization of Two Research Laboratories. In *Proceedings of the ACM Conference on Computer-Supported Cooperative Work, CSCW '92*, 330–337. New York: Association for Computing Machinery.

Harper, R., ed. 2003. *Inside the Smart Home.* London: Springer-Verlag.

Hartsock, N. 1983. The Feminist Standpoint: Developing the Ground for a Specifically Feminist Historical Materialism. In *Discovering Reality*, ed. S. Harding and M. B. Hintikka, 283–310. Dordrecht: Kluwer.

Hartsock, N. 1998. *The Feminist Standpoint Revisited and Other Essays.* Boulder, CO: Westview Press.

Harvey, D. 1989. *The Condition of Postmodernity: An Enquiry into the Origins of Cultural Change.* Oxford: Blackwell.

Harvey, D. 2001. *Spaces of Capital: Towards a Critical Geography.* London: Routledge.

Hasbrouck, J., and S. Faulkner. 2006. "Why Are You Taking My Picture?" Navigating the Cultural Contexts of Visual Procurement. In *Proceedings of the Ethnographic Praxis in Industry Conference 2006.*

Hayes, G. R., L. Gardere, G. D. Abowd, and K. N. Truong. 2008. CareLog: A Selective Archiving Tool for Behavior Management in Schools. In *CHI 2008: Proceedings of the ACM Conference on Human Factors in Computing Systems*, 685–694. New York: Association for Computing Machinery.

Hayles, K. 1999. *How We Became Posthuman: Virtual Bodies in Cybernetics, Literature, and Informatics.* Chicago: University of Chicago Press.

Heath, C., and P. Luff. 1992. Collaboration and Control: Crisis Management and Multimedia Technology in London Underground Control Rooms. *Computer-Supported Cooperative Work* 1: 69–94.

Heath, C., P. Luff, D. vom Lehn, J. Hindmarsh, and J. Cleverly. 2002. Crafting Participation: Designing Ecologies, Configuring Experience. *Visual Communication* 1 (1): 9–33.

Helmreich, S. 1998. *Silicon Second Nature: Culturing Artificial Life in a Digital World.* Berkeley: University of California Press.

Heng, G., and J. Devan. 1992. State Fatherhood: The Politics of Nationalism, Sexuality, and Race in Singapore. In *Nationalisms and Sexualities*, ed. A. Parker, M. Russo, D. Somer, and P. Yaeger, 343–364. New York: Routledge.

Herdt, G. H. 1994. *Guardians of the Flutes: Idioms of Masculinity, with a New Preface.* 2nd ed. Chicago: University of Chicago Press.

Hill, R. 1995. Blackfellas and Whitefellas: Aboriginal Land Rights, the Mabo Decision, and the Meaning of Land. *Human Rights Quarterly* 17 (2): 303–322.

Hindmarsh, J., C. Heath, D. vom Lehn, and J. Cleverly. 2002. Creating Assemblies: Aboard the Ghost Ship. In *Proceedings of the ACM Conference on Computer-Supported*

Cooperative Work, CSCW 2002, 156–165. New York: Association for Computing Machinery.

Hindus, D., S. Mainwaring, N. Leduc, A. E. Hagstrom, and O. Bayley. 2001. Casablanca: Designing Social Communication Devices for the Home. In *CHI 2001: Proceedings of the ACM Conference on Human Factors in Computing Systems*, 325–332. New York: Association for Computing Machinery.

Hofstede, G. 2001. *Culture's Consequences: Comparing Values, Behaviors, Institutions, and Organizations across Nations*. 2nd ed. Newbury Park, CA: Sage.

Hong, J., G. Boriello, J. Landay, D. McDonald, B. Schilit, and J. Tygar. 2003. Privacy and Security in the Location-Enhanced World Wide Web. Paper presented at Ubicomp 2003 Workshop on Privacy and Security in Ubiquitous Computing.

Hornecker, E., and J. Buur. 2006. Getting a Grip on Tangible Interaction: A Framework on Physical Space and Social Interaction. In CHI '06: *Proceedings of the ACM Conference on Human Factors in Computing Systems, 437–446*. New York: Association for Computing Machinery.

Horowitz, I. L., ed. 1967. *The Rise and Fall of Project Camelot: Studies in the Relationship between Social Science and Practical Politics*. Cambridge, MA: MIT Press.

Horrigan, B. 1986. The Home of Tomorrow, 1927–1945. In *Imagining Tomorrow: History, Technology, and the American Future*, ed. J. J. Corn, 137–163. Cambridge, MA: MIT Press.

Housewife's Garden Shed Porn. 2000. BBC News. Available at <http://news.bbc.co.uk/2/hi/uk_news/wales/811643.stm>.

Hsu, F. L. K. 1953. *Americans and Chinese: Two Ways of Life*. New York: Schuman.

Hsu, F. L. K. 1963. *Clan, Caste, and Club*. Princeton, NJ: Van Nostrand Reinhold.

Hsu, F. L. K. 1983. *Rugged Individualism Reconsidered: Essays in Psychological Anthropology*. Knoxville: University of Tennessee Press.

Huff, W. G. 1994. *The Economic Growth of Singapore: Trade and Development in the 20th Century*. Cambridge: Cambridge University Press.

Hughes, J., D. Randall, and D. Shapiro. 1993. From Ethnographic Record to Systems Design: Some Experiences from the Field. *Computer-Supported Cooperative Work* 1: 123–141.

Hui, O. J., T. C. Kiong, and T. E. Ser, eds. 1997. *Understanding Singapore Society*. Singapore: Times Academic Press.

Hull, R., B. Clayton, and T. Melamed. 2004. Rapid Authoring of Mediascapes. In *Proceedings of the International Conference on Ubiquitous Computing, Ubicomp 2004*, ed. N. Davies, E. Mynatt, and I. Siio,125–142. Berlin: Springer.

Humble, J., A. Crabtree, T. Hemmings, K.-P. Åkesson, B. Koleva, T. Rodden, and P. Hansson. 2003. Playing with the Bits: User Configuration of Ubiquitous Domestic Environments. In *Proceedings of the International Conference on Ubiquitous Computing, Ubicomp 2003*, ed. A. Dey, A. Schmidt, and J. McCarthy, 256–263.

Hutchins, E. 1995. *Cognition in the Wild.* Cambridge, MA: MIT Press.

Hutchinson, H., H. Hansen, N. Roussel, B. Eiderbäck, W. Mackay, B. Westerlund, B. Bederson, et al. 2003. Technology Probes: Inspiring Design for and with Families. In *CHI '03: Proceedings of the ACM Conference on Human Factors in Computing Systems*, 17–24. New York: Association for Computing Machinery.

Iachello, G., I. Smith, S. Consolvo, G. D. Abowd, J. Hughes, and F. Howard, F. Potter, J. Scott, T. Sohn, J. Hightower, and A. LaMarca. 2005a. Control, Deception, and Communication: Evaluating the Deployment of a Location-Enhanced Messaging Service. In *Proceedings of the International Conference on Ubiquitous Computing, UbiComp 2005*, ed. M. Beigl, S. Intille, J. Rekimoto, and H. Tokuda, 213–231.

Iachello, G., I. Smith, S. Consolvo, M. Chen, and G. Abowd. 2005b. Developing Privacy Guidelines for Social Location Disclosure Applications and Services. In *Proceedings of the Symposium on Usable Privacy and Security, SOUPS 2005*, 65–76. New York: Association for Computing Machinery.

IDA (Infocomm Development Authority of Singapore). 2006. The iN2015 Masterplan. Available at <http://www.ida.gov.sg/About%20us/20070907161958.aspx>.

IDA (Infocomm Development Authority of Singapore). 2010a. Annual Survey of Infocomm Usage in Households for 2009. Available at <http://www.ida.gov.sg/doc/Publications/Publications_Level3/Survey2009/HH2009ES.pdf> (accessed June 2010).

IDA (Infocomm Development Authority of Singapore). 2010b. Stats on Telecom Services for 2010 Jan–Jun. Available at <http://www.ida.gov.sg/Publications/20100205111257.aspx> (accessed June 2010).

Igoe, T. 2007. *Making Things Talk: Practical Methods for Connecting Physical Objects.* Sebastopol, CA: O'Reilly.

IST Advisory Group. 2001. *Scenarios for Ambient Intelligence in 2010.* Seville: Institute for Prospective Technological Studies.

Ito, M., and D. Okabe. 2005a. Intimate Connections: Contextualizing Japanese Youth and Mobile Messaging. In *The Inside Text: Social Perspectives on SMS in the Mobile Age*, ed. R. Harper, L. Palen, and A. Taylor, 127–146. Dordrecht: Kluwer.

Ito, M., and D. Okabe. 2005b. Technosocial Situations: Emergent Structurings of Mobile Email Use. In *Personal, Portable, Pedestrian: Mobile Phones in Japanese Life*, ed. M. Ito, D. Okabe, and M. Matsuda, 257–273. Cambridge, MA: MIT Press.

Ito, M., D. Okabe, and M. Matsuda. 2005. *Personal, Portable, Pedestrian: Mobile Phones in Japanese Life*. Cambridge, MA: MIT Press.

Jenkins, H. 1992. *Textual Poachers: Television Fans and Participatory Culture*. London: Routledge.

Jones, G. 2004. *Shed Men*. New Holland Publishers.

Ju, W., R. Hurwitz, T. Judd, and B. Lee. 2001. CounterActive: An Interactive Cookbook for the Kitchen Counter. In *CHI '01: Proceedings of the ACM Conference on Human Factors in Computing Systems*, 269–270. New York: Association for Computing Machinery.

Jungnickel, K. 2004. Seventy-three Urban Journeys. Workshop on ubiquitous computing on the urban frontier, Ubicomp 2004, Nottingham, UK.

Jungnickel, K. 2007. *DIY WiFi: Making Homebrew High-Technology in Suburban Backyards* exhibition. Adelaide, Australia.

Jungnickel, K. 2008. Disorderly Design: The Role of Mess in the Making of WiFi. Paper presented at the annual meeting of the Society for the Social Studies of Science, Rotterdam.

Jungnickel, K. 2009. *"Makers," "Mashers," and "Mods": Grassroots Technology Practices in Australian Backyard Culture Exhibition*. London: Goldsmiths.

Jungnickel, K., and G. Bell. 2008. Home Is Where the Hub Is? Wireless Infrastructures and the Nature of Domestic Culture in Australia. In *Urban Informatics: The Practice and Promise of the Real-time City*, ed. M. Foth. Hershey, PA: Information Science Reference, IGI Global.

Kahneman, D., and A. Tversky. 1979. Prospect Theory: An Analysis of Decision under Risk. *Econometrica* 47 (2): 263–292.

Kayan, S., S. R. Fussell, and L. D. Setlock. 2006. Cultural Differences in the Use of Instant Messaging in Asia and North America. In *Proceedings of the 2006 Twentieth Anniversary Conference on Computer-Supported Cooperative Work*, 525–528. New York: Association for Computing Machinery.

Keesing, R. 1974. Theories of Culture. *Annual Review of Anthropology* 3: 73–97.

Kelleher, W. 2003. *The Troubles in Ballybogoin: Memory and Identity in Northern Ireland*. Ann Arbor: University of Michigan Press.

Kelley, K., and H. Francis. 1993. Places Important to Navajo People. *American Indian Quarterly* 17 (2): 151–169.

Kidd, C. D., R. Orr, G. D. Abowd, C. G. Atkeson, I. A. Essa, B. MacIntyre, E. D. Mynatt, T. Starner, and W. Newstetter. 1999. The Aware Home: A Living Laboratory for Ubiquitous Computing Research. In *Proceedings of the Second International*

Workshop on Cooperative Buildings, Integrating Information, Organization, and Architecture. London: Springer-Verlag.

Kim, C. S. 1986. *The Culture of Korean Industry: An Ethnography of Poongsan Corporation*. Tuscon: University of Arizona Press.

Kim, E. M. 1997. *Big Business, Strong State: Collusion and Conflict in South Korean Development, 1960–1990*. New York: State University of New York Press.

Kindberg, T., A. Sellen, and E. Geelhoed. 2004. Security and Trust in Mobile Interactions: A Study of Users' Perceptions and Reasoning. In *Proceedings of the International Conference on Ubiquitous Computing, Ubicomp 2004*, ed. N. Davies, E. Mynatt, and I. Siio, 196–213. Berlin: Springer.

King, A. D. 1984. *The Bungalow: The Production of a Global Culture*. London: Routlege and Kegan Paul.

Klein, N. 1997. *The History of Forgetting: Los Angeles and the Erasure of Memory*. London: Verso.

Kleinman, S., and G. Fine. 1979. Rhetorics and Action in Moral Organizations: Social Control of Little Leaguers and Ministry Students. *Urban Life* 8 (3): 275–294.

Kling, R. 1994. Reading "All about" Computerization: How Genre Conventions Shape Non-Fiction Social Analysis. *Information Society* 10 (3): 147–172.

Kling, R., G. McKim, J. Fortuna, and A. King. 2000. Scientific Collaboratories as Socio-Technical Interaction Networks: A Theoretical Approach. Paper presented at the American Conference on Information Systems, Long Beach, CA.

Kluckhohn, C., and W. H. Kelly. 1945. The Concept of Culture. In *The Science of Man in the World Culture*, ed. R. Linton, 78–105. New York: Octagon Books.

Knauft, B. 1996. *Genealogies of the Present in Cultural Anthropology*. New York: Routledge.

Koh, G., and G. L. Ooi. 2000. *State-Society Relations in Singapore*. Oxford: Oxford University Press.

Koo, H., ed. 1993. *State and Society in Contemporary Korea*. Ithaca, NY: Cornell University Press.

Kroeber, A. L., and Kluckhohn, C. 1952. Culture: A Critical Review of Concepts and Definitions. Harvard University Peabody Museum of American Archeology and Ethnology Papers 47.

Lally, E. 2002. *At Home with Computers*. Oxford: Berg.

Lamming, M., and M. Flynn. 1994. Forget-me-not: Intimate Computing in Support of Human Memory. In *Proceedings of the International Symposium on Next-Generation Human Interface FRIEND21*.

Landzelius, K. 2007. *Native on the Net: Indigenous Cyber-Activism and Virtual Diasporas over the World Wide Web*. London: Routledge.

Langley, W. 2005. Women Are Spending too Much Time in Sheds, and not Everyone's Happy about It. Available at <http://www.telegraph.co.uk/news/main.jhtml?xml=/news/2005/10/09/nshed09.xml>.

Larsen, E. 2002. *Wired Churches, Wired Temples: Taking Congregations and Missions into Cyberspace*. Pew Internet Research Organization. Available at <http://www.pewinternet.org/report_display.asp?r=28>.

Latour, B. 1987. *Science in Action: How to Follow Scientists and Engineers through Society*. Cambridge, MA: Harvard University Press.

Latour, B. 2005. *Reassembling the Social: An Introduction to Actor-Network Theory*. Oxford: Oxford University Press.

Lave, J., and E. Wenger. 1991. *Situated Learning: Legitimate Peripheral Participation*. Cambridge: Cambridge University Press.

Law, J. 1987. Technology and Heterogeneous Engineering: The Case of Portuguese Expansion. In *The Social Construction of Technological Systems*, ed. W. Bijker, T. Hughes, and T. Pinch, 111–134. Cambridge, MA: MIT Press.

Lawson. H. 1900. A Rough Shed. In *On the Track*. London: Angus and Robertson. Available at <http://setis.library.usyd.edu.au/ozlit/pdf/p00019.pdf>.

Leary, S. Men's Sheds. 2005. Australian Broadcasting Corporation, Western Plains, New South Wales. Available at<http://www.abc.net.au/westernplains/stories/s1375040.htm>.

LeFebvre, H. 1991. *The Production of Space*. Trans. D. Nicholson-Smith. Oxford: Blackwell.

Lett, D. P. 1998. *Pursuit of Status: The Making of South Korea's "New" Urban Middle Class*. Cambridge, MA: Harvard University Press.

Leunig, M. 1990. *The Traveling Leunig*. Sydney: Penguin.

Lévi-Strauss, C. 1969. *The Elementary Structures of Kinship*. Boston: Beacon Press.

Lévi-Strauss, C. 1970. *The Raw and the Cooked*. Trans. J. Weightman and D. Weightman. New York: Harper and Row.

Ling, R., and B. Yttri. 2002. Hyper-Coordination via Mobile Phones in Norway. In *Perpetual Contact: Mobile Communication, Private Talk, Public Performance*, ed. J. E. Katz and M. Aakhus. Cambridge: Cambridge University Press.

Livingstone, S. 1992. The Meaning of Domestic Technologies: A Personal Construct Analysis of Familial Gender Relations. In *Consuming Technology: Media and Information in Domestic Spaces*, ed. R. Silverstone and E. Hirsch, 113–130. London: Routledge.

Livingstone, S. 2002. *Young People, New Media: Childhood and the Changing New Media Environment.* Newbury Park, CA: Sage.

Lock, M. 1998. Breast Cancer: Reading the Omens. *Anthropology Today* 14: 8–16.

Long, S., D. Aust, G. D. Abowd, and C. Atkeson. 2006. *Cyberguide: Prototyping Context-Aware Mobile Applications.*

Low, S. 2000. *On the Plaza: The Politics of Public Space and Culture.* University of Texas Press.

Lu, S., and G. A. Fine. 1995. The Presentation of Ethnic Authenticity: Chinese Food as a Social Accomplishment. *Sociological Quarterly* 36 (3): 535–553.

Lutz, C. 1986. Emotion, Thought, and Estrangement: Emotion as a Cultural Category. *Cultural Anthropology* 1 (3): 287–309.

Lutz, C. 1988. *Unnatural Emotions: Everyday Sentiments on a Micronesian Atoll and Their Challenge to Western Theory.* Chicago: University of Chicago Press.

Lynch, K. 1960. *The Image of the City.* Cambridge, MA: MIT Press.

Lyons, M. 2003. Spatial Segregation in Seven Cities: A Longitudinal Study of Home Ownership, 1971–91. *Housing Studies* 18 (3): 305–326.

Mackay, W. 1990. Patterns of Sharing Customizable Software. In *Proceedings of the ACM Conference on Computer-Supported Cooperative Work, CSCW'90,* 209–221. New York: Association for Computing Machinery.

MacLean, A., K. Carter, L. Lövstrand, and T. Moran. 1990. User-Tailorable Systems: Pressing the Issues with Buttons. In *CHI '90: Proceedings of the ACM Conference on Human Factors in Computing Systems,* 175–182. New York: Association for Computing Machinery.

Madian, A. L., and A. N. Oppenheim. 1969. Knowledge for What? The Camelot Legacy: The Dangers of Sponsored Research in the Social Sciences. *British Journal of Sociology* 20 (3): 326–336.

Mainwaring, S., K. Anderson, and M. Chang. 2005. Living for the Global City: Mobile Kits, Urban Interfaces, and Ubicomp. In *Proceedings of the International Conference on Ubiquitous Computing, Ubicomp 2005,* ed. M. Beigl, S. Intille, J. Rekimoto, and H. Tokuda, 269–286. Berlin: Springer.

Mainwaring, S., M. Chang, and K. Anderson. 2004. Infrastructures and Their Discontents. In *Proceedings of the International Conference on Ubiquitous Computing, Ubicomp 2004,* Ed. N. Davies, E. Mynatt, and I. Siio, 418–432. Berlin: Springer.

Mainwaring, S. D., and A. Woodruff. 2005. Investigating Mobility, Technology, and Space in Homes, Starting with "Great Rooms." In *Proceedings of EPIC 2005,* 188–195. Berkeley: University of California Press.

Malinowski, B. 1922. *Argonauts of the Western Pacific*. New York: E. P. Dutton and Co. Inc.

Malkki, L. 1992. National Geographic: The Rootedness of Peoples and the Territorialization of National Identity amongst Scholars and Refugees. *Cultural Anthropology* 7 (1): 24–44.

Malkki, L. 1995. *Purity and Exile: Memory and National Cosmology amongst Hutu Refugees in Tanzania*. Chicago: University of Chicago Press.

March, W., and C. Fleuroit. 2006. Girls, Technology, and Privacy: "Is My Mother Listening?" In *CHI '06: Proceedings of the ACM Conference on Human Factors in Computing Systems*, 107–110. New York: Association for Computing Machinery.

Marcus, A. 2001. Cross-cultural User-Interface Design for Work, Home, Play, and on the Way. In *Proceedings of the Nineteenth Annual International Conference on Computer Documentation*, 221–222. New York: Association for Computing Machinery.

Marcus, G. 1995. Ethnography in/of the World System: The Emergence of Multi-sited Ethnography. *Annual Review of Anthropology* 24: 95–117.

Marcus, G., and M. Fischer. 1986. *Anthropology as Cultural Critique: An Experimental Moment in the Social Sciences*. Chicago: University of Chicago Press.

Marston, S. 2000. The Social Construction of Scale. *Progress in Human Geography* 24: 219–242.

Martin, J. 1996. *The Dialectical Imagination: A History of the Frankfurt School and Institute of Social Research, 1923–1950*. Berkeley: University of California Press.

Marx, K. 1859/1977. *A Contribution to the Critique of Political Economy*. Repr., Moscow: Progress Publishers.

Massey, A. P., Y. C. Hung, M. Montoya-Weiss, and V. Ramesh. 2001. When Culture and Style Aren't about Clothes: Perceptions of Task-Technology "Fit" in Global Virtual Teams. In *Proceedings of the 2001 International ACM SIGGROUP Conference on Supporting Group Work*, ed. C. Ellis and I. Zigurs, 207–213. New York: Association for Computing Machinery.

Massey, D. 1993. Power-Geometry and a Progressive Sense of Place. In *Mapping the Futures: Local Cultures, Global Change*, ed. J. Bird, B. Curtis, T. Putnam, G. Robertson, and L. Tickner, 60–70. London: Routledge.

Masui, T., K. Tsukada, and I. Siio. 2004. MouseField: A Simple and Versatile Input Device for Ubiquitous Computing. In *Proceedings of the International Conference on Ubiquitous Computing, Ubicomp 2004*, 319–328. Berlin: Springer.

McCullough, M. 2004. *On Digital Ground: Architecture, Pervasive Computing, and Environmental Knowing*. Cambridge, MA: MIT Press.

Meggitt, M. 1962. *Desert People: A Study of the Walbiri Aborigines of Australia*. Chicago: University of Chicago Press.

Menon, A. P. G. 2000. ERP in Singapore: A Perspective a Year On. *Tec* 41 (2).

Menon, A. P. G., and C. K. Keong. 1998. The Making of Singapore's Electronic Road Pricing System. In *Proceedings of the International Conference on Transportation into the Next Millennium*.

Merten, D. 1999. Enculturation into Secrecy among Junior High School Girls. *Journal of Contemporary Ethnography* 28 (2): 107–137.

MIC (Ministry of Information and Communication). 2004. *IT 839 Strategy: The Road to $20,000 GDP/Capita*. Korea: Ministry of Information and Communication.

MIC (Ministry of Information and Communication). 2005. *IT 839: A Leap to Advanced Korea Based on IT*. Korea: Ministry of Information and Communication. Available at <http://www.mic.go.kr/www.DynamicITKorea.org>.

Michaels, E. 1986. *The Aboriginal Invention of Television in Central Australia, 1982–1985*. Canberra: Australian Institute of Aboriginal Studies.

Michaels, E. 1987. *For a Cultural Future: Frances Jupurrurla Makes TV at Yuendumu*. Malvern, UK: Artspace.

Miller, D. 2003. The Virtual Moment. *Journal of the Royal Anthropological Institute* 9 (1): 57–75.

Miller, D. 2009. *The Comfort of Things*. Cambridge, UK: Polity.

Miller, D., and D. Slater. 2000. *The Internet: An Ethnographic Approach*. New York: New York University Press.

Moran, T. P., and R. J. Anderson. 1990. The Workaday World as a Paradigm for CSCW Design. In *Proceedings of the 1990 ACM Conference on Computer-Supportive Cooperative Work*, 381–393. New York: Association for Computing Machinery.

Morgan, L. H. 1851/2006. *The League of the Ho-de-no-sau-nee or Iroquois*. Whitefish, MT: Kessinger Publishing.

Mosco, V. 2004. *The Digital Sublime: Myth, Power, and Cyberspace*. Cambridge, MA: MIT Press.

Munn, N. 1996. Excluded Spaces: The Figure in the Australian Aboriginal Landscape. *Critical Inquiry* 22: 446–465.

Murray, G., and A. Perera. 1996. *The Global City State*. Singapore: China Library.

Myers, F. 1979. Emotions and the Self: A Theory of Personhood and Political Order amongst Pintupi Aboriginals. *Ethos* 7 (4): 343–370.

Myers, F. 1986. *Pintupi Country, Pintupi Self: Sentiment, Place, and Politics amongst Western Desert Aborigines*. Washington, DC: Smithsonian Institute Press.

Nader, L., ed. 1996. *Naked Science: Anthropological Inquiry into Boundaries, Power, Knowledge*. New York: Routledge.

Nagel, T. 1986. *The View from Nowhere*. Oxford: Oxford University Press.

Nardi, B. 1993. *A Small Matter of Programming: Perspectives on End-User Computing*. Cambridge, MA: MIT Press.

National Computer Board. 1992. *A Vision of an Intelligent Island: IT2000 Report*. Singapore: National Computer Board.

Nelson, L. 2000. *Measured Excess, Status, Gender, and Consumer Nationalism in South Korea*. New York: Columbia University Press.

Neumann, P. 1995. *Computer-Related Risks*. Upper Saddle River, NJ: Addison-Wesley.

Newman, W. M., M. A. Eldridge, and M. Lamming. 1991. PEPYS: Generating Autobiographies by Automatic Tracking. In *Proceedings of the Second European Conference on Computer-Supported Work*, ed. L. Bannon, M. Robinson, and K. Schmidt. 175–189. Dordrecht: Kluwer.

NIAC (National Internet Advisory Committee). 2004. *Report of the National Advisory Council, 2003*. Singapore: Ministry of Information, Communications, and the Arts.

NIDA (National Internet Development Agency). 2004. Survey of the Computer and Internet Usage, Ministry for Information and Communication, National Internet Development Agency of Korea. Available at <http://isis.nida.or.kr/eng_report_down/upload/user_sum_200412_english.pdf>.

NIDA (National Internet Development Agency). 2007. Survey of the Computer and Internet Usage, Ministry for Information and Communication, National Internet Development Agency of Korea. Available at <http://isis.nida.or.kr/eng_report_down/upload/user_sum_200412_english.pdf>.

Nippert-Eng, C. 1996. *Home and Work: Negotiating Boundaries in Everyday Life*. Chicago: University of Chicago Press.

Nippert-Eng, C. 2005a. Boundary Play. *Space and Culture* 8 (3): 302–324.

Nippert-Eng, C. 2005b. Usable Privacy. Paper presented at the Usable Privacy Conference, Intel Corporation, Hillsboro, OR.

Nippert-Eng, C. 2006. Privacy by Design. Paper presented at the Design Research Society International Conference, Lisbon.

Nippert-Eng, C. 2007. Privacy in the United States: Some Implications for Design. *International Journal of Design* 1 (2): 1–14.

Nippert-Eng, C., and J. Melican. 2004. Disclosure and Concealment: Wallets, Purses, and Identity Work in Modern Societies. Paper presented at the annual meeting of the American Sociological Association, San Francisco.

Nippert-Eng, C., J. Melican, R. Hinman, and R. Pikkel. 2005. Social Accessibility and ICTs: New Challenges to an Old Problem. Paper presented at the annual meeting of the American Sociological Association, Philadelphia.

Norman, D. 2004. *Emotional Design: Why We Love (or Hate) Everyday Things*. New York: Basic Books.

Norrie, M., and B. Signer. 2005. Overlaying Paper Maps with Digital Information Services for Tourists. In *Proceedings of the ENTER 2005 Conference on Travel and Tourism Technology*, 23–33.

O'Neill, E., V. Kostakos, T. Kindberg, A. Scheick, A. Penn, D. Fraser, and T. Jones. 2006. Instrumenting the City: Developing Methods for Observing and Understanding the Digital Cityscape. In *Proceedings of the International Conference on Ubiquitous Computing, Ubicomp 2006*, ed. P. Dourish and A. Friday, 315–332. Berlin: Springer.

Ong, W. 1988. *Orality and Literacy: The Technologizing of the Word*. London: Routledge.

Orr, J. 1996. *Talking about Machines: An Ethnography of a Modern Job*. Ithaca, NY: Cornell University Press.

Ortner, S. 1984. Theory in Anthropology since the Sixties. *Comparative Studies in Society and History* 26 (1): 126–166.

Palen, L., M. Salzman, and E. Youngs. 2000. Going Wireless: Behavior and Practice of New Mobile Phone Users. In *Proceedings of the Conference on Computer-Supported Cooperative Work, CSCW'00*, 201–210. New York: Association for Computing Machinery.

Parthasarathy, S. 2004. Regulating Risk: Defining Genetic Privacy in the United States and Britain. *Science, Technology, and Human Values* 29 (3): 332–352.

Patel, S., and G. Abowd. 2004. The ContextCam: Augmented Point of Capture Video Annotation. In *Proceedings of the International Conference on Ubiquitous Computing, Ubicomp 2004*, ed. N. Davies, E. Mynatt, and I. Siio, 301–318. Berlin: Springer.

Patel, S., J. Kientz, G. Hayes, S. Bhat, and G. Abowd. 2006. Farther Than You May Think: An Empirical Investigation of the Proximity of Users to Their Mobile Phones. In *Proceedings of the International Conference on Ubiqutious Computing, Ubicomp 2006*, ed. P. Dourish and A. Friday, 123–140.

Patil, S., and J. Lai. 2005. Who Gets to Know What When: Configuring Privacy Preferences in an Awareness Application. In *CHI '05: Proceedings of the ACM Conference in Human Factors in Computing Systems*, 101–110. New York: Association for Computing Machinery.

Paulos, E., and T. Jenkins. 2005. Urban Probes: Encountering Our Emerging Urban Atmospheres. In *CHI '05: Proceedings of the ACM Conference on Human Factors in Computing Systems,* 341–350. New York: Association for Computing Machinery.

Pearce, J. M., J. Murphy, and W. Smith. 2008. Supporting Gardeners to Plan Domestic Watering: A Case Study of Designing an "Everyday Simulation." In *Proceedings of the Australasian Conference on Computer-Human Interaction, OzCHI 2008*, 227–230.

Pelto, P. 1973. *Snowmobile Revolution: Technology and Social Change in the Arctic.* Upper Saddle River, NJ: Addison-Wesley.

Penley, C. 1997. *NASA/Trek: Popular Science and Sex in America.* New York: Verso.

Picard, R. 1997. *Affective Computing.* Cambridge, MA: MIT Press.

Pickering, A. 2010. *The Cybernetic Brain: Sketches of Another Future.* Chicago: University of Chicago Press.

Povinelli, E. 1993. Might Be Something: The Language of Indeterminacy in Aboriginal Land Use. *Man* 28: 679–704.

Prøitz, L. 2005. Intimacy Fiction: Intimate Discourses in Mobile Telephony Communication amongst Young Norwegian People. In *A Sense of Place: The Global and the Local in Mobile Communication,* ed. N. Kristof. Vienna: Passagen Verlag.

Pullman-Jones, S. 2005. Using Photographic Data to Build a Large-scale Global Comparative Visual Ethnography of Domestic Spaces: Can a Limited Data Set Capture the Complexities of "Sociality"? In *Ethnographic Praxis in Industry Conference Proceedings,* 128–139.

Rabin, M. 1998. Psychology and Economics. *Journal of Economic Literature* 36: 11–46.

Reeves, S., M. Fraser, H. Schnadelbach, C. O'Malley, and S. Benford. 2005. Engaging Augmented Reality in Public Places. In *CHI '05: Adjunct roceedings of the ACM Conference on Human Factors in Computing Systems.* New York: Association for Computing Machinery.

Rengier, P. 1987. *City-State in South-East Asia.* Singapore: University of Hawaii Press.

Richardson, L. 1988. Secrecy and Status: The Social Construction of Forbidden Relationships. *American Sociological Review* 53 (2): 209–219.

Richtel, M. 2004. For Liars and Loafers, Cellphones Offer an Alibi. *New York Times,* June 26.

Riva, G., F. Vatalaro, F. Davide, and M. Alcañiz, eds. 2005. *Ambient Intelligence: The Evolution of Technology, Communication, and Cognition towards the Future of Human-Computer Interaction.* Amsterdam: IOS Press.

Robison, R., and D. S. G. Goodman, eds. 1996. *The New Rich in Asia: Mobile Phones, McDonald's, and Middle-Class Revolution*. London: Routledge.

Rodden, T., and S. Benford. 2003. The Evolution of Buildings and Implications for the Design of Ubiquitous Domestic Environments. In *Proceedings of the ACM Conference on Human Factors in Computing Systems, CHI 2003*, 9–16. New York: Association for Computing Machinery.

Rodden, T., A. Crabtree, T. Hemmings, B. Koleva, J. Humble, K.-P. Åkesson, and P. Hansson. 2004. Between the Dazzle of a New Building and Its Eventual Corpse: Assembling the Ubiquitous Home. In *Proceedings of the ACM Conference on Designing Interactive Systems*, 71–80. New York: Association for Computing Machinery.

Rode, J., E. Toye, and A. Blackwell. 2004. The Fuzzy Felt Ethnography: Understanding the Programming Patterns of Domestic Appliances. *Personal and Ubiquitous Computing* 8: 161–176.

Rogers, Y. 1997. Reconfiguring the Social Scientist: Shifting from Telling Designers What to Do to Getting More Involved. In *Social Science, Technical Systems, and Cooperative Work: Beyond the Great Divide*, ed. G. Bowker, L. Star, and B. Turner, 57–78. Mahwah, NJ: Lawrence Erlbaum.

Rogers, Y. 2006. Moving on from Weiser's Vision of Calm Computing: Engaging Ubicomp Experiences. In *Proceedings of the Eighth International Conference on Ubiquitous Computing, UbiComp 2006*, ed. P. Dourish and A. Friday, 404–421. New York: Springer.

Rosaldo, R. 1983. The Shame of Headhunters and the Autonomy of Self. *Ethos* 11 (3): 135–151.

Rosner, D. 2010 Mediated Crafts: Digital Practices around Creative Handwork. In *Proceedings of the Twenty-eighth International Conferences on Human Factors in Computing Systems, Extended Abstracts*, 2955–2958.

Rosner, D., and J. Bean. 2009. Learning from IKEA Hacking: "I'm not One to Decoupage a Tabletop and Call It a Day. In CHI '09: *Proceedings of the ACM Conference on Human Factors in Computing Systems*, 419–422. New York: Association for Computing Machinery.

Rosner, D. K., and K. Ryokai. 2008. Spyn: Augmenting Knitting to Support Storytelling and Reflection. In *Ubicomp '08: Proceedings of the Tenth International Conference on Ubiquitous Computing*, 340–349.

Rosner, D. K., and K. Ryokai. 2009. Probing the Creative Process of Everyday Knitters. In *Proceedings of the Conference on Creativity and Cognition*, 195–204. New York: Association for Computing Machinery.

Royal Society for the Prevention of Accidents. 2002. Home and Leisure Accident Surveillance System: Annual Report. Available at <http://www.hassandlass.org.uk/query/reports.htm>.

Rudolph, L. 2001. Project Oxygen: Pervasive, Human-Centric Computing—An Initial Experience. In *Proceedings of the Advanced Information Systems Engineering: Thirteenth International Conference*, ed. K. R. Dittrich, A. Geppert, and M. C. Norrie. New York: Springer.

Rydell, R. 1993. *World of Fairs: The Century-of-Progress Expositions*. Chicago: University of Chicago Press.

Sahlins, M. 1972. *Culture and Practical Reason*. Chicago: University of Chicago Press.

Satchell, C., and P. Dourish. 2009. Beyond the User: Use and Non-use in HCI. In *Proceedings of the Australasian Conference on Computer-Human Interaction, OzCHI 2009*, 9–16.

Satyanarayanan, M. 2005. Swiss Army Knife or Wallet? *IEEE Pervasive Computing Magazine* 4 (2): 2–3.

Schegloff, E. 1972. Formulating Place: Notes on a Conversational Practice. In *Studies in Social Interaction*, ed. D. Sudnow, 75–119. New York: Free Press.

Scheter, B. 2000. Seeing the Light: IBM's Vision of Life beyond the PC. *IBM Research* 2.

Schilit, B., A. LaMarca, G. Borriello, W. Griswold, D. McDonald, E. Lazowska, A. Balachandran, J. Hong, and V. Iverson. 2003. Challenge: Ubiquitous Location-Aware Computing and the "Place Lab" Initiative. In *Proceedings of the ACM International Workshop on Wireless Mobile Applications and Services on WLAN*, 29–35.

Schmidt, K. 2000. The Critical Role of Workplace Studies in CSCW. In *Workplace Studies: Rediscovering Work Practice and Informing Design*, ed. P. Luff, J. Hindmarsh, and C. Heath, 141–149. Cambridge: Cambridge University Press.

Schneider, D. M. 1980. *American Kinship: A Cultural Account*. Chicago: University of Chicago Press.

Schneier, B. 2000. *Secrets and Lies: Digital Security in a Networked World*. New York: Wiley.

Schutz, A. 1943. The Problem of Rationality in the Social World. *Economica* 10 (38): 130–149.

Scott, J. C. 1985. *Weapons of the Weak: Everyday Forms of Peasant Resistance*. New Haven, CT: Yale University Press.

Scott, J. C. 1998. *Seeing Like a State: How Certain Schemes to Improve the Human Condition Have Failed*. New Haven, CT: Yale University Press.

Scott, J. C. 2009. *The Art of Not Being Governed: An Anarchist History of Upland Southeast Asia*. New Haven, CT: Yale University Press.

Seo, K. W. 2006. The Law of Conservation of Activities in Domestic Spaces. *Journal of Asian Architecture and Building Engineering* 5 (1): 21–28.

Setlock, L. D., S. R. Fussell, and C. Neuwirth. 2004. Taking It Out of Context: Collaborating within and across Cultures in Face-to-face Settings and via Instant Messaging. In *Proceedings of the 2004 ACM Conference on Computer-Supported Cooperative Work*, 604–613. New York: Association for Computing Machinery.

Shapin, S. 2003. How to Eat Like a Gentleman: Dietetics and Ethics in Early Modern England. In *Right Living: An Anglo-American Tradition of Self-help Medicine and Hygiene*, ed. C. E. Rosenberg, 21–58. Baltimore: Johns Hopkins University Press.

Shapin, S. 2004. The Great Neurotic Art. *London Review of Books* 26 (15): 16–18.

Sharp, L. 1952. Steel Axes for Stone-Age Australians. *Human Organization* 11 (2): 17–22.

Sherman, W. 1995. *John Dee: The Politics of Reading and Writing in the English Renaissance*. Amherst: University of Massachusetts Press.

Sherry, J., and T. Salvador. 2001. Running and Grimacing: The Struggle for Balance in Mobile Work. In *Wireless World: Social, Cultural, and Interactional Issues in Mobile Communications and Computing*, ed. B. F. Brown, N. Green, and R. Harper. 108–120. London: Springer-Verlag.

Shirky, C. 2008. *Here Comes Everybody: The Power of Organizing with Organizations*. New York: Penguin Press.

Shklovski, I., J. Vertesi, E. Troshynski, and P. Dourish. 2009. The Commodification of Location: Dynamics of Power in Location-Based Systems. In *Ubicomp '09: Proceedings of the Eleventh International Conference on Ubiquitous Computing*, 11–20. New York: Association for Computing Machinery.

Shklovsky, V. 1925/1990. *Theory of Prose* [O теорий прозы, O teorii prozy]. Trans. B. Sher. Naperville, IL: Dalkey Archive.

Silverstone, R., and E. Hirsch, eds. 1992. *Consuming Technology: Media and Information in Domestic Spaces*. London: Routledge.

Slaughter, S., T. Campbell, M. Holleman, and E. Morgan. 2002. The "Traffic" in Graduate Students: Graduate Students as Tokens of Exchange between Academe and Industry. *Science, Technology, and Human Values* 27 (2): 282–312.

Smail, D. L. 1999. *Imaginary Cartographies: Possession and Identity in Late Medieval Marseille*. Ithaca, NY: Cornell University Press.

Smith, B. C. 1996. *On the Origin of Objects*. Cambridge, MA: MIT Press.

Smith, I., S. Consolvo, A. LaMarca, J. Hightower, J. Scott, T. Sohn, J. Huges, G. Iachello, and G. D. Abowd. 2005. Social Disclosure of Place: From Location Technology to Communication Practices. In *Proceedings of Third International Conference on Pervasive Computing. Eds. H.-W. Gellerson, A. Schmidt, and R. Want*, 134–151. London: Springer-Verlag.

Smith, V., ed. 1989. *Hosts and Guests: The Anthropology of Tourism*. Pittsburgh, PA: University of Pennsylvania Press.

Spiekermann, S., J. Grossklags, and B. Berendt. 2001. E-privacy in Second Generation E-commerce: Privacy Preferences versus Actual Behavior. In *Proceedings of the ACM Conference on Electronic Commerce, EC '01*, 38–47. New York: Association for Computing Machinery.

Spradley, J. 1988. *You Owe Yourself a Drunk: An Ethnography of Urban Nomads*. New York: University Press of America.

Sproull, L., and S. Kiesler. 1991. *Connections: New Ways of Working in the Networked Organization*. Cambridge, MA: MIT Press.

Stanner, W. E. H. 1958. The Dreaming. In *Reader in Comparative Religion*, ed. E. Z. Vogt and W. A. Lessa, 158–167. New York: Harper and Row.

Star, S. L. 1999. The Ethnography of Infrastructure. *American Behavioral Scientist* 43 (3): 377–391.

Star, S. L. 2002. Infrastructure and Ethnographic Practices. *Scandinavian Journal of Information Systems* 24 (2): 107–122.

Star, S. L., and J. Griesemer. 1989. Institutional Ecology, "Translations," and Boundary Objects: Amateurs and Professionals in Berkeley's Museum of Vertebrate Zoology, 1907–39. *Social Studies of Science* 19 (3): 387–420.

Star, S. L., and K. Ruhleder. 1994. Steps Towards an Ecology of Infrastructure: Complex Problems in Design and Access for Large-scale Collaborative Systems. In *Proceedings of the ACM Conference on Computer-Supported Cooperative Work, CSCW'94*, 253–264. New York: Association for Computing Machinery.

Stewart, J., S. Bauman, M. Escobar, J. Hilden, K. Bihani, and M. W. Newman. 2008. Accessible Contextual Information for Urban Orientation. In *Ubicomp '08: Proceedings of the Tenth International Conference on Ubiquitous Computing*, 332–335. New York: Association for Computing Machinery.

Sterling, B. 2005. *Shaping Things*. Cambridge, MA: MIT Press.

Stocking, G. W., Jr. 1982. *Race, Culture, and Evolution: Essays in the History of Anthropology*. Chicago: University of Chicago Press.

Stocking, G. W., Jr., ed. 1985. *Observers Observed: Essays in Ethnographic Fieldwork*. Madison: University of Wisconsin Press.

Stocking, G. W., Jr. 1991. *Victorian Anthropology*. New York: Free Press.

Stocking, G. W., Jr. 1992. *The Ethnographer's Magic and Other Essays in the History of Anthropology*. Madison: University of Wisconsin Press.

Stocking, G. W., Jr. 1998. *After Tylor: British Social Anthropology, 1888–1951*. Madison: University of Wisconsin Press.

Stoffle, R., D. Halmo, and D. Austin. 1997. Cultural Landscapes and Traditional Cultural Properties: A Southern Paiute View of the Grand Canyon and Colorado River. *American Indian Quarterly* 21 (2): 229–249.

Strathern, M. 1988. *The Gender of the Gift*. Berkeley: University of California Press.

Strathern, M., ed. 2000. *Audit Cultures: Anthropological Studies in Accountability, Ethics, and the Academy*. London: Routledge.

Strathern, M. 2004. *Commons and Borderlands: Working Papers on Interdisciplinarity, Accountability, and the Flow of Knowledge*. Wantage, UK: Sean Kingston Publishing.

Strausser, S. 1982. *Never Done: A History of American Housework*. New York: Pantheon.

Strong, R., and W. Gaver. 1996. Feather, Scent, and Shaker: Supporting Simple Intimacy. In *Videos, Demos, and Short Papers of CSCW '96*, 29–30.

Sturgeon, W. 2004. British Are a Nation of Cowardly Texters. Available at <http://networks.silicon.com/mobile/0,39024665,39117896,00.htm> (accessed October 18, 2008).

Suchman, L. 1987. *Plans and Situated Actions: The Problem of Human-Machine Communication*. Cambridge: Cambridge University Press.

Suchman, L. 1999. Introductory Remarks. *Work Practices and Technology Fest* (May). Available at <http://www.workpractice.com/wpt-fest/introduction.html>.

Suchman, L. 2002. Located Accountabilities in Technology Production. *Scandinavian Journal of Information Systems* 14: 91–105.

Suchman, L. 2007. *Human-Machine Reconfigurations: Plans and Situated Actions*. 2nd ed. New York: Cambridge University Press.

Taylor, C. 2002. Modern Social Imaginaries. *Public Culture* 14 (1): 91–124.

Tennenhouse, D. 2000. Proactive Computing. *Communications of the ACM* 43 (5): 43–50.

Thompson, E. P. 1967. Time, Work-Discipline, and Industrial Capitalism. *Past and Present* 38: 56–97.

Thomson, M. 2002. *The Complete Blokes and Sheds: Stories from the Shed*. London: Angus and Robertson.

Thorburn, G. 2002. *Mens and Sheds*. New Holland Publishers.

Tolmie, P., J. Pycock, T. Diggins, A. MacLean, and A. Karsenty. 2001. Unremarkable Computing. In *CHI '01: Proceedings of the ACM Conference on Human Factors in Computing Systems*, 399–406. New York: Association for Computing Machinery.

Tran, Q. T., G. Calcaterra, and E. Mynatt. D. 2005. Cook's Collage: Déjà vu Display for a Home Kitchen. In *Proceedings of HOIT, Home-Oriented Informatics and Telematics Conference,* 15–32.

Trautmann, T. R. 1988. *Lewis Henry Morgan and the Invention of Kinship.* Berkeley: University of California Press.

Traweek, S. 1988. *Beamtimes and Lifetimes: The World of High-Energy Physicists.* Cambridge, MA: Harvard University Press.

Troshynski, E., C. Lee, and P. Dourish. 2008. Accountabilities of Presence: Reframing Location-Based Systems. In *CHI '08: Proceedings of the ACM Conference on Human Factors in Computing Systems,* 487–496. New York: Association for Computing Machinery.

Truong, K., E. Huang, and G. Abowd. 2004. CAMP: A Magnetic Poetry Interface for End-User Programming of Capture Applications for the Home. In *Proceedings of the International Conference on Ubiquitous Computing, Ubicomp 2004,* ed. N. Davies, E. Mynatt, and I. Siio,143–160. Berlin: Springer.

Turnbull, D. 2000. *Masons, Tricksters, and Cartographers: Comparative Studies in the Sociology of Scientific and Indigenous Knowledge.* Amsterdam: Harwood.

Turner, F. 2006. *From Counterculture to Cyberculture: Steward Brand, the Whole Earth Network, and the Rise of Digital Utopianism.* Chicago: University of Chicago Press.

Turner, G. 2002. *British Cultural Studies: An Introduction.* 3rd ed. London: Routledge.

Turner, P., and E. Davenport, eds. 2005. *Spaces, Spatiality, and Technology.* Berlin: Springer.

Urry, J. 2000. *Sociology beyond Societies: Mobilities for the Twenty-first Century.* London: Routledge.

Urry, J. 2007. *Mobilities.* Cambridge, UK: Polity.

Van Gennep, A. 1960. *Rites of Passage.* Chicago: University of Chicago Press.

Vermuri, S., C. Schmandt, W. Bender, S. Teller, and B. Lassey. 2004. An Audio-Based Personal Memory Aid. In *Proceedings of the International Conference on Ubiquitous Computing, Ubicomp 2004,* ed. N. Davies, E. Mynatt, and I. Siio, 400–417. Berlin: Springer.

Verran, H. 1998. Re-imagining Land Ownership in Australia. *Postcolonial Studies* 1 (2): 237–254.

Verrips, J., and B. Meyer. 2001. Kwaku's Car: The Struggles and Stories of a Ghanian Long-distance Taxi-Driver. In *Car Cultures,* ed. D. Miller, 153–184. Oxford: Berg.

Vertesi, J. 2008. Mind the Gap: The London Underground Map and Users' Representations of Urban Space. *Social Studies of Science* 38 (1): 7–33.

vom Lehn, D., J. Hindmarsh, and C. Heath. 2001. Exhibiting Interaction: Conduct and Collaboration in Museums and Galleries. *Symbolic Interaction* 24: 189–216.

Want, R., A. Hooper, V. Falcão, and J. Gibbons. 1992. The Active Badge Location System. *ACM Transactions on Information Systems* 10 (1): 91–102.

Want, R., B. N. Schilit, N. I. Adams, R. Gold, K. Petersen, D. Goldberg, J. R. Ellis, and M. Weiser. 1995. The ParcTab Ubiquitous Computing Experiment. Xerox PARC Technical Report CSL-95-1 (March).

Ward, G. 2005. Rat Sparks NZ Telecoms Blackout. BBC News Online. Available at <http://news.bbc.co.uk/2/hi/asia-pacific/4114556.stm> (accessed October 21, 2008).

Webster, F. 2004. Cultural Studies and Sociology at, and after, the Closure of the Birmingham School. *Cultural Studies* 18 (6): 8447–8462.

Weick, K. 1995. *Sensemaking in Organizations*. Newbury Park, CA: Sage.

Weinberg, N. 1996. Competition Calling. *Forbes* 158 (11): 146–147.

Weirich, D., and A. Sasse. 2001. Pretty Good Persuasion: Steps Towards Effective Password Security in the Real World. In Proceedings of the ACM New Security Paradigms Workshop, 137–143.

Weiser, M. 1991. The Computer for the 21st Century. *Scientific American* 265 (3): 94–104.

Weiser, M. 1993. Hot Topics: Ubiquitous Computing. *IEEE Computer* (October).

Weiser, M. 1994. Building Invisible Interface? Keynote talk at the ACM Symposium on User Interface, Systems, and Technologies. Available at <http://www.ubiq.com/hypertext/weiser/UbiHome.html> (accessed June 2010).

Weiser, M., R. Gold, and J. S. Brown. 1999. Origins of Ubiquitous Computing Research at PARC in the Late 1980s. *IBM Systems Journal* 38 (4): 693–696.

Weiser, M., and J. Seely Brown. 1997. The Coming Age of Calm Computing. In *Beyond Calculation: The Next Fifty Years of Computing*, ed. P. Denning and R. Metcalfe. New York: Copernicus.

Wiggershaus, R. 1995. *The Frankfurt School: Its History, Theories, and Political Significance*. Trans. M. Robertson. Cambridge, MA: MIT Press.

Williams, A., P. Dourish, and K. Anderson. 2008. Anchored Mobilities: Mobile Technology and Transnational Migration. In *Proceedings of the ACM Conference on Designing Interactive Systems, DIS 2008*, 323–332. New York: Association for Computing Machinery.

Wilson, G. 2005. *Shed Heaven*. BBC online newsmagazine. Available at <http://news.bbc.co.uk/2/hi/uk_news/magazine/4543675.stm> (accessed June 2010).

Wolfe, P. 1991. On Being Woken Up: The Dreamtime in Anthropology and in Australian Settler Culture. *Comparative Studies in Society and History* 33 (2): 197–224.

Wood, D. 1992. *The Power of Maps*. New York: Guilford Press.

Woolgar, S., ed. 2002. *Virtual Society? Technology, Cyperbole, Reality*. Oxford: Oxford University Press.

Wright, Edgar, dir. 2004. *Shaun of the Dead*. London: Big Talk Productions.

Wyche, S., P. Sengers, and R. E. Grinter. 2006. Historical Analysis: Using the Past to Design the Future. In *Proceedings of the Eighth International Conference on Ubiquitous Computing, Ubicomp '06*, ed. P. Dourish and A. Friday, 35–51. Berlin: Springer.

Wyche, S., A. Taylor, and J. Kaye. 2007. Pottering: A Design-Oriented Investigation. In *CHI '07: Proceedings of the ACM Conference on Human Factors in Computing Systems*, 1893–1898. New York: Association for Computing Machinery.

Wynne, B. 1992. Misunderstood Misunderstandings: Social Identities and Public Uptake of Science. *Public Understanding of Science* 1 (3): 281–304.

Yates, J. 1993. *Control through Communication: The Rise of System in American Management*. Baltimore: Johns Hopkins University Press.

Yengoyan, A. 1986. Theory in Anthropology: The Demise of the Concept of Culture. *Comparative Studies in Society and History* 28 (2): 368–374.

Yew, L. K. 2000. *From Third World to First: The Singapore Story, 1965–2000*. New York: HarperCollins.

Yoon, Soon Young. 1979. Women's Studies in Korea. *Signs: Journal of Women in Culture and Society* 4 (4): 751–762.

Zelkha, E., and B. Epstein. 1998. From Devices to "Ambient Intelligence." Paper presented at the Digital Living Room conference, June.

Zerubavel, E. 1991. *The Fine Line: Making Distinctions in Everyday Life*. New York: Free Press.

Index

Aboriginal Australia, 26, 80–82, 104–106, 142
Abowd, Gregory, 16
Abu-Lughod, Lila, 76, 78–79
Ackerman, Mark, 67, 73
Active Badge, 137
Actor-Network Theory, 134
Adams, Anne, 151
Affect and interaction design, 75–79
Agre, Philip, 40, 205
Ambient intelligence, 15
Anderson, Ken, 26
Anderson, Robert, 67–70
Anthropology, social and cultural, 46
Apache, 126
Appadurai, Arjun, 84
ARPANET, 50
Art galleries, empirical studies, 103
Asokan, Ashwini, 146
Audit culture, 196
Austin, Diane, 199
Australian Broadcasting Corporation (ABC), 168
Australia, 167–170, 172, 182. *See also* Aboriginal Australia

Barth, Frederik, 196
Basso, Keith, 126
Baudrillard, Jean, 164
Bazerman, Charles, 65
Beck, Ulrich, 153
Beijing, 139
Benedict, Ruth, 48
Bestor, Theodore, 130
Blackwell, Alan, 182–183
Blast Theory, 121
Bluetooth, 109
Boas, Franz, 47
Boehner, Kirsten, 79
Bowker, Geoffrey, 27
Brown, John Seely, 10
Burundi, 82–83
Büscher, Monika, 130
Bush, Vannevar, 17
Button, Graham, 67

Calm computing, 18, 25
Can You See Me Now?, 121
Center for Contemporary Cultural Studies, 55
Chang, Michele, 26
Chernobyl disaster, 152
Chicago School, 48
China, People's Republic of, 103, 139
Collaborative virtual environment (CVE), 99
Computer-supported cooperative work (CSCW), 72
Coyne, Richard, 205
Critical standpoint theory, 170, 205
Critical theory, 54–55, 205
Cultural landscapes, 199

Culture
oral and literate, 196
taxonomic and generative accounts, 51–54
Cybernetics, 4

Day, Kristen, 152–153, 175–176
de Certeau, Michel, 123
de Laet, Marianne, 134
Dee, John, 196
Defamiliarization, 130
Diet, 151
Domestic violence, 177, 182
Douglas, Mary, 151
Dramatic computer, 91
Dreaming (Aboriginal Australia), 80–82, 104–106

Economics (metaphor), 143
Edwards, Keith, 173, 207
Emotion. *See* Affect and interaction design
Equator IRC, 15
Ethnography and theory, 68
Ethnomethodology, 49
Everquest, 108

Faulkner, Susan, 150
Fine, Gary Alan, 155
Fischer, Michael, 130
Forsythe, Diana, 70
Foucault, Michel, 145
Frankfurt School, 54
Frederick, Christine, 166

Geertz, Clifford, 51, 61
Georgia Institute of Technology, 15, 16, 63
Gibson, William, 41
Giddens, Anthony, 153
Gilbreth, Frank, 166
Gilbreth, Lillian, 166
Goffman, Erving, 82
Goodwin, Charles, 101, 130
Goodwin, Marjory Harkness, 130
Goody, Jack, 196
Goss, Jon, 164
GPS (global positioning by satellite), 83, 109, 121, 136, 139, 148, 155, 157
Graham, Stephen, 97
Gramsci, Antonio, 55
Greenfield, Adam, 138, 188
Grinter, Rebecca, 173, 207
GSM (global system for mobile communications), 131

Halmo, David, 199
Harvey, David, 122
Heath, Christian, 101
Hofstede, Gerd, 52
Honeywell Kitchen Computer, 161–163, 177
Hutchins, Ed, 196
Hutu, 82–83

India, 110–111, 140–141
Interdisciplinarity, 62, 190
International Classification of Diseases (ICD), 27–28
Internet protocol, 50, 99
Inuit, 136
Irvine, CA, 175
IT2000 (Singapore), 31–32
IT839 (Korea), 38
Ito, Mizuko, 112, 132, 193

Jungnickel, Katrina, 102

Kaiditch, 104–106
Kay, Alan, 2, 12
Kelleher, William, 129
Kilburn High Road, 42
Kling, Rob, 50
Korea, 29, 36

Lancaster University, 63, 85
Lash, Scott, 153
Latour, Bruno, 23
Leunig, Michael, 168
Lévi-Strauss, Claude, 69
Ling, Richard, 132
London Underground, 83, 102, 122
Luff, Paul, 101
Lutz, Catherine, 76–78
Lying online, 147–149
Lynch, Kevin, 193

Mainwaring, Scott, 26
Malinowski, Bronislaw, 47, 181
Malkki, Liisa, 82
Maori, 107
Marcus, George, 56, 88, 130
Marseille, 128
Marvin, Simon, 97
Massey, Doreen, 42, 129, 194
McCarthy, John, 205
McCullough, Malcolm, 26, 188
McPherson, Tara, 189, 191
Mecca, 100
Mediascapes, 84
Melikan, Jay, 150
Merten, Don, 154
Metro. *See* Paris Metro
Meyer, Birgit, 27
Migration, 119
Miller, Daniel, 56, 71–72, 199
Mobile telephony, 97, 111, 148, 156
Mol, Annemarie, 134
Moore's Law, 12
Mosco, Vincent, 4
Multi-sited ethnography, 56, 82, 88
Munn, Nancy, 80–82
Mynatt, Elizabeth, 16

Native American culture, 126, 199
Navigation, 196
New Zealand, 107, 112
Nippert-Eng, Christena, 110, 146
Northern Ireland, 129

Occasion maps, 196
Okabe, Daisuke, 112, 132, 193
Olivetti Research Laboratory, 137
Ong, Walter, 196

Paiute, 199
PARC. *See* Xerox Palo Alto Research Center
Panopticon, 145, 193
Papua New Guinea, 47, 53. *See also* Trobriand Islands
Paris Metro, 100
Parole monitoring, 157–158
Pepys (computer system), 138
Pervasive computing, 15
Pickering, Andrew, 5
Place Lab (computer system), 143
Polycropping, 194
Pornography, 141, 142, 179
Power geometries, 42, 129, 194
Professional vision, 101
Project Camelot, 48
Proximate future, 20, 23

Radio 4 (BBC), 168
Rank Xerox EuroPARC, 137
Reflexive modernization, 153
Religion and technology, 111
RFID (radio frequency identification), 143, 156, 176
Risk society, 153
Rode, Jennifer, 182–183
Rogers, Yvonne, 18

Sahlins, Marshall, 159
Sal scenario, 12–13, 117
Sasse, Martina Angela, 151
Schegloff, Emanuel, 101
Schmidt, Kjeld, 72
Schneier, Bruce, 160

Schütte-Lihotzky, Margarete, 166
Scott, James, 119, 193–194
Seamful design, 43, 110
Second Life, 108
Sengers, Phoebe, 205
Shapin, Steven, 151
Shaun of the Dead, 175
Shed, 167
Singapore, 31
 and censorship, 35
Slater, Don, 56, 71–72
Smaill, Daniel Lord, 128
Smart home, 164
Smith, Brian Cantwell, 62
SMS (short message service), 33
Southern Paiute, 199
Star, Susan Leigh, 26, 27, 95, 96, 115
Sterling, Bruce, 188
Stoffle, Richard, 199
Strathern, Marilyn, 53, 199
Subincision, 145
Suchman, Lucy, 10, 17
Symbolic interactionism, 47

Tanzania, 82–83
Taylor, Frederick, 166
Thailand, 140
Theory object, 189, 191
Thompson, E. P., 166
Thomson, Mark, 167
Time-space compression, 122
Tokyo, 132
Tolmie, Peter, 91
Toye, Eleanor, 182–183
Treaty of Waitangi, 107
Trinidad, 56–57, 71–72
Trobriand Islands, 47, 88

Ubiquitous Dream Hall, 29–30
Ultima Online, 108
Undersound, 83
University of California, Berkeley, 138
University of Chicago, 46
Urban informatics, 97, 123
Urry, John, 119

Verrips, Jojada, 27
Vertesi, Janet, 122, 129
Virtualism, 196,

Warlpiri, 80–82, 104–106, 145
Weick, Karl, 79
Weiser, Mark, 1, 9, 95
Western Apache, 126
WiFi, 109, 115, 121, 131
Wildavsky, Aaron, 151
Wireless spectrum rights, 107
World of Warcraft, 108
Wright, Peter, 205
Wynne, Brian, 152

Xerox Palo Alto Research Center (PARC), 1, 9, 63

Yttri, Birgitte, 132